Mangroves as Fish Habitat

Edited by

KAREN J. MURCHIE
Department of Biology, College of The Bahamas
Post Office Box F-42766, Freeport, Grand Bahama, The Bahamas

PEDRAM P. DANESHGAR
Marine and Environmental Biology and Policy Program, Monmouth University
400 Cedar Avenue, West Long Branch, New Jersey 07764, USA

American Fisheries Society Symposium 83

Proceedings of the *2nd International Symposium on Mangroves as Fish Habitat*
Held in Mazatlan, Mexico
April 7–12, 2014

American Fisheries Society
Bethesda, Maryland, USA
2015

Suggested citation formats follow.

Entire book

Murchie, K. J., and P. P. Daneshgar, editors. 2015. Mangroves as fish habitat. American Fisheries Society, Symposium 83, Bethesda, Maryland.

Chapter in book

Barreto, C. R., P. P. Daneshgar, and J. A. Tiedemann. 2015. Carbon sequestration of dwarf red mangrove in The Bahamas. Pages 101–112 *in* K. J. Murchie and P. P. Daneshgar, editors. Mangroves as fish habitat. American Fisheries Society, Symposium 83, Bethesda, Maryland.

Front cover: artwork by Bruno Amezcua.

Printed in the United States of America on acid-free paper.

Library of Congress Control Number 2015942746
ISBN 978-1-934874-42-4
ISSN 0892-2284

American Fisheries Society Web site: www.fisheries.org

American Fisheries Society
5410 Grosvenor Lane, Suite 110
Bethesda, Maryland 20814
USA

Mangroves as Fish Habitat

Contents

Preface

In April 2006, the *First International Symposium on Mangroves as Fish Habitat* was held at the University of Miami's Rosenstiel School of Marine and Atmospheric Science, and in May 2007 the proceedings were published in a special issue of the *Bulletin of Marine Science*, edited by Joseph E. Serafy and Rafael J. Araújo. At the time of the first symposium, scientists and conservation groups had only recently begun delving into mangrove–fisheries relationships. Techniques such as stable isotope analyses, otolith microchemistry, and biotelemetry were used to gain insight into a variety of topics from diel and seasonal fish movements in mangroves (i.e., for feeding, refugia, nursery function) to the importance of mangrove connectivity to other habitats. Also included in the discussion was the emerging field of mangrove habitat restoration.

A few years later, scientists from Ecologists Without Borders, the American Fisheries Society, and the Fisheries Conservation Foundation started planning for a second symposium. The steering committee consisted of Dave Philipp and Julie Claussen of the Fisheries Conservation Foundation, Eric Knudsen of Ecologists Without Borders, Cleve Steward of the Sustainable Fisheries Foundation, John Tiedemann of Monmouth University, and Felipe Amezcua of Universidad Nacional Autónoma de México. Because significant advances in mangrove–fisheries linkages had been made since that first symposium, the *2nd International Symposium on Mangroves as Fish Habitat* was held in April of 2014 at the El Cid Convention Center in Mazatlan, Mexico in conjunction with the joint annual meeting of the Western Division and the Mexican Chapter of the American Fisheries Society. Drawing over 150 scientists, fisheries managers, and students representing 13 countries, the symposium consisted of 35 oral presentations organized around five major themes, each thematic session starting with a keynote speaker. Before the symposium even began, Eric Carey of the Bahamas National Trust provided an excellent and stimulating Western Division plenary address related to the role of mangroves in recreational fisheries. J. Eugenio Barrios of World Wildlife Fund México was the keynote speaker for the theme "ecology of mangroves and their fisheries." The session on "climate change and the mangrove ecosystem" enjoyed a keynote address by Boone Kaufman of Oregon State University. The "conservation and restoration of mangrove ecosystems" theme and the "habitat loss of mangroves" theme were led by keynote addresses from Francisco Flores Verdugo of the Universidad Nacional Autónoma de México and Stacey Lubin-Gray of the Bahamas BEST Commission, respectively. At the end of the three-day symposium, various scientists working on mangrove-fishery issues around the globe participated in a workshop led by James Hutchison of the University of Cambridge to parameterize a preliminary qualitative model of the spatial distribution of benefits mangroves provide to fisheries. The resultant manuscript is included in this publication.

While investigations into basic ecological functions of mangrove systems were still at the core of the discussions during the symposium and make up the bulk of the manuscripts and extended abstracts in these proceedings, a marked shift in the urgency of assigning dollar values to mangrove fisheries is noticeable. Also, the concept of using recreational fisheries as a conservation tool for mangrove habitats is presented as we continue to work towards protecting these crucial ecosystems. Mangroves are iconic features in subtropical and tropical coastal systems worldwide, and their own diversity along with the variety of life forms they support require the scientific community to continue to advocate for their protection and, where already lost, their restoration. It will be interesting to see how the main themes for research change with the third international symposium on mangroves as fish habitat—we hope that there will be more success stories to be shared and that the overall decline in mangrove habitat loss will have been further slowed or, even better, eliminated.

Several key sponsors are to be recognized for making the symposium possible. They include the Western Division of the American Fisheries Society, the Fisheries Conservation Foundation, Ambata Capital Partners, Monmouth University, the Bahamas National Trust, Bonefish & Tarpon Trust, the Cape Eleuthera Institute, the Estuaries Section of the American Fisheries Society, Ecologists Without Borders, the Sustainable Fisheries Foundation, the Illinois Natural History Survey, the Department of Natural Resources and Environmental Sciences at the University of Illinois, and the Universidad Nacional Autónoma de México. A number of individuals helped in various ways to make the symposium a success and include Eric Carey, Aaron Schultz, Michael Philipp, Pam Sponholtz, Hilda Sexauer, Aaron Adams, and James Hutchison.

For the publication of the *2nd International Mangroves as Fish Habitat* proceedings, we need to thank a number of key people. First, the gorgeous cover art is courtesy of Bruno Amezcua. Second, thank you to Debby Lehman and Aaron Lerner for their work and support at making this publication possible. Last, we are grateful to the 34 reviewers listed below who provided excellent feedback to the authors, ensuring the quality and integrity of the research you are about to enjoy.

Felipe Amezcua, Universidad Nacional Autónoma de México, Mexico
Rich Bastian, Monmouth University, USA
Ross Boucek, Florida International University, USA
Jake Brownscombe, Carleton University, Canada
Kusmana Cecep, Bogor Agricultural University, Indonesia
Jocelyn Curtis-Quick, Cape Eleuthera Institute, The Bahamas
Sascha Danylchuk, Fish Mission, USA
Dennis DeVries, Auburn University, USA
Kerry Flaherty, Florida Fish and Wildlife Conservation Commission, USA
Luis Flores, Escuela Superior Politecnica del Litoral, Ecuador
Chandri Giri, US Geological Survey, USA
Kathy Guindon, Florida Fish and Wildlife Conservation Commission, USA
Don Hagen, Clemson University, USA
Eric Holzmueller, Southern Illinois University, USA
Catherine Jadot, ES Caribbean, Turks and Caicos Islands
Muhammad Yamin Janjua, Fisheries and Oceans Canada, Canada
Timothy Jardine, University of Saskatchewan, Canada
Monica Lara, St Petersburg College, USA
Craig Layman, Florida International University, USA
Shing Lip Lee, Griffith University, Australia
Tiffany Medley, Monmouth University, USA
Vivian Nguyen, Carleton University, Canada
Amanda O'Toole, Pacific Northwest National Laboratory, USA
Linwood Pendleton, Duke University, USA
Dave Philipp, University of Illinois, USA
Wes Porak, Florida Fish and Wildlife Conservation Commission, USA

Jennifer Rehage, Florida International University, USA
Krista Sherman, University of Exeter, United Kingdom
Geoffrey Smith, University of Florida, USA
Jeff Stein, University of Illinois, USA
Philip Stevens, Florida Fish and Wildlife Conservation Commission, USA
Brendan Talwar, Florida State University, USA
John Tiedemann, Monmouth University, USA
Brian Wilsey, Iowa State University, USA

Symbols and Abbreviations

The following symbols and abbreviations may be found in this book without definition. Also undefined are standard mathematical and statistical symbols given in most dictionaries.

A	ampere
AC	alternating current
Bq	becquerel
C	coulomb
°C	degrees Celsius
cal	calorie
cd	candela
cm	centimeter
Co.	Company
Corp.	Corporation
cov	covariance
DC	direct current; District of Columbia
D	dextro (as a prefix)
d	day
d	dextrorotatory
df	degrees of freedom
dL	deciliter
E	east
E	expected value
e	base of natural logarithm (2.71828...)
e.g.	(exempli gratia) for example
eq	equivalent
et al.	(et alii) and others
etc.	et cetera
eV	electron volt
F	filial generation; Farad
°F	degrees Fahrenheit
fc	footcandle (0.0929 lx)
ft	foot (30.5 cm)
ft^3/s	cubic feet per second ($0.0283\ m^3/s$)
g	gram
G	giga (10^9, as a prefix)
gal	gallon (3.79 L)
Gy	gray
h	hour
ha	hectare (2.47 acres)
hp	horsepower (746 W)
Hz	hertz
in	inch (2.54 cm)
Inc.	Incorporated
i.e.	(id est) that is
IU	international unit
J	joule
K	Kelvin (degrees above absolute zero)
k	kilo (10^3, as a prefix)
kg	kilogram
km	kilometer
l	levorotatory
L	levo (as a prefix)
L	liter (0.264 gal, 1.06 qt)
lb	pound (0.454 kg, 454g)
lm	lumen
log	logarithm
Ltd.	Limited
M	mega (10^6, as a prefix); molar (as a suffix or by itself)
m	meter (as a suffix or by itself); milli (10^{-3}, as a prefix)
mi	mile (1.61 km)
min	minute
mol	mole
N	normal (for chemistry); north (for geography); newton
N	sample size
NS	not significant
n	ploidy; nanno (10^{-9}, as a prefix)
o	ortho (as a chemical prefix)
oz	ounce (28.4 g)
P	probability
p	para (as a chemical prefix)
p	pico (10^{-12}, as a prefix)
Pa	pascal
pH	negative log of hydrogen ion activity
ppm	parts per million

qt quart (0.946 L)
R multiple correlation or regression coefficient
r simple correlation or regression coefficient
rad radian
S siemens (for electrical conductance); south (for geography)
SD standard deviation
SE standard error
s second
T tesla
tris tris(hydroxymethyl)-aminomethane (a buffer)
UK United Kingdom
U.S. United States (adjective)
USA United States of America (noun)
V volt
V, Var variance (population)
var variance (sample)
W watt (for power); west (for geography)
Wb weber
yd yard (0.914 m, 91.4 cm)
α probability of type I error (false rejection of null hypothesis)
β probability of type II error (false acceptance of null hypothesis)
Ω ohm
μ micro (10^{-6}, as a prefix)
′ minute (angular)
″ second (angular)
° degree (temperature as a prefix, angular as a suffix)
% per cent (per hundred)
‰ per mille (per thousand)

Mangrove–Fishery Linkages

American Fisheries Society Symposium 83:3–15, 2015

The Current State of Knowledge on Mangrove Fishery Values

James Hutchison* and Philine zu Ermgassen
Department of Zoology, University of Cambridge
Downing Street, Cambridge CB2 3EJ, UK

Mark Spalding
Department of Zoology, University of Cambridge
Downing Street, Cambridge, CB2 3EJ, UK
and
Global Marine Team, The Nature Conservancy
Department of Zoology, University of Cambridge
Downing Street, Cambridge CB2 3EJ, UK

Abstract.—Mangroves are widely understood to be important habitats for fisheries, supporting resident fish, crustacean, and mollusk populations as well as acting as nursery grounds for species that are targeted by offshore fisheries. There is, however, a lack of quantitative data on fisheries that operate in and around mangroves. We carried out a systematic search to gather data on mangrove fisheries from the scientific literature. We filtered the 4,358 studies returned by the search based on their title and abstract and extracted data from 169 of these. Despite the abundance of literature on mangrove fisheries, we were unable to build a data set of comparable, quantitative data of sufficient size to support numerical modeling approaches. In part, this is due to the variety of mangrove fisheries, which range from small-scale subsistence fishing for mollusks and crabs to large-scale industrialized prawn trawling. This is compounded by the broad range of reporting methods and metrics encountered in the literature. We make a number of recommendations to guide the future reporting of mangrove fisheries to allow for better quantification and comparison of fisheries values at large spatial scales.

Introduction

Situated in tropical, low energy, and primarily estuarine environments, mangroves provide a suite of benefits to fish and invertebrates, including enhanced primary productivity and refuge from predation (Laegdsgaard and Johnson 2001; Ewe et al. 2006). While it is widely accepted that mangroves play a key role in enhancing fisheries productivity, the absolute degree of enhancement as well as the temporal and spatial variation in mangrove fisheries productivity are poorly resolved. Mangroves are well studied and there is an abundance of literature on mangrove supported fisheries, so this apparent lack of quantitative evidence at a large scale arises primarily from the many disparate ways in which mangrove fisheries are reported.

One key reason for the diverse ways in which mangrove fisheries are reported is

* Corresponding author: jtwh3@cam.ac.uk

that mangrove fisheries are themselves extremely diverse. Mangrove fisheries range from highly specific crab and mollusk fisheries through small-scale mixed species fisheries to large-scale offshore commercial fisheries for species that use mangroves at earlier stages of their life cycle. This range of scales has an accompanying variety of fishing methods and gear types. Mangrove forests themselves are largely impenetrable to large boats or machinery so most fisheries within the mangrove are small scale. Small-scale fisheries are frequently underreported in government statistics and in the literature (FAO and World Fish Centre 2008; Costello et al. 2012). Much of the catch never enters formal markets, being consumed by fishers, their families, and the local community. Fishing may not even be a full-time occupation of many of those involved (Coates 2002). The small-scale, artisanal nature of many in-mangrove fisheries therefore presents an additional challenge to gathering consistent fisheries data.

Finally, there is considerable uncertainty over the extent of the benefits that mangroves provide to offshore fisheries. Several studies have shown correlations between fishery catches and mangrove area, but this alone does not necessarily imply causality (Manson et al. 2005). Such correlations can therefore not be used to determine the extent to which mangroves enhance fishery catches above the background level provided by other estuarine and coastal habitats. Gaining a true insight into the value of mangroves to fisheries at a large scale is therefore a significant challenge.

To gain a comprehensive overview of the current state of knowledge of mangrove fisheries values and to support a longer-term effort to model and map mangrove fisheries values (see Hutchison et al. 2015, this volume), we carried out a systematic literature search. We used a standardized, repeatable search methodology to gather literature on mangrove fisheries from online literature databases. Here, we describe the results of this literature search along with the initial results of a synthesis of some of this material.

Methods

Literature review

We followed a standardized systematic literature search process (based on Collaboration for Environmental Evidence 2013), which enabled an iterative approach of testing and refining search terms to improve both comprehensiveness and specificity. Search terms were combined into sets and searches run to return only papers with at least one term from each set. The title, abstract, and keywords of each article were searched using Web of Knowledge (www.webofknowledge.com), recording the total number of results returned. The title and abstract of the first 100 results of a search were scanned to assess the proportion of those results that were potentially relevant.

We began with two sets of search terms, one relating to mangroves and the other to fishery target groups (Table 1). The initial search returned a large number of results of which only a small percentage were relevant. Changing the term "fish*" to "fisher*" removed a number of papers that included the word "fish" but had no relevance to fisheries. However, overall relevance remained low. Removing the mangrove genera from the mangrove term list gave a significant improvement in relevance while yielding nearly as many papers.

Many of the papers focused on fish biology and had little relevance to fisheries, so we added a third set of terms related to the action of catching or harvesting fish and invertebrates (e.g., "yield," "harvest," and "catch") or to the function of mangroves in enhancing fish populations (e.g., "survival," "nursery," and "growth"). This yielded a similar number of results but with much greater relevance. The three sets of search terms were

Table 1.—The initial sets of search terms used, and the number and relevance of the results returned using each set in Web of Knowledge. * indicates a wildcard operator such that "fisher*" would also include fisheries, fishermen, and so forth.

Mangrove terms	Fish terms	Results	Relevance
Acanthus, Acrostichum, Aegialitis, Aegiceras, Aglaia, Avicennia, Bruguiera, Camptostemon, Ceriops, Cynometra, Diospyros, Dolichandrone, Excoecaria, Heritiera, Kandelia, Lumnitzera, mangrove, *Nypa, Osbornia, Pelliciera, Pemphis, Rhizophora, Scyphiphora, Sonneratia, Tabebuia, Xylocarpus*	cockle*, crab*, finfish, fish*, oyster*, prawn*, shrimp*	5,186	16/100
Acanthus, Acrostichum, Aegialitis, Aegiceras, Aglaia, Avicennia, Bruguiera, Camptostemon, Ceriops, Cynometra, Diospyros, Dolichandrone, Excoecaria, Heritiera, Kandelia, Lumnitzera, mangrove, *Nypa, Osbornia, Pelliciera, Pemphis, Rhizophora, Scyphiphora, Sonneratia, Tabebuia, Xylocarpus*	cockle*, crab*, finfish, fisher*, oyster*, prawn*, shrimp*	3,596	18/100
mangrove*	cockle*, crab*, finfish, fisher*, oyster*, prawn*, shrimp*	3,432	28/100

therefore used for all further searches. Final fine-tuning included the addition of "valu*" (value, valuation, etc.) to capture some missing economic valuation papers. "Fisher*" was also changed back to "fish*" in the fish terms, as the action terms filtered out papers with no fishery relevance. The final set of search terms used is shown in Table 2.

Once the final set of search terms had been established, it was also used to search Scopus (www.scopus.com) and Science Direct (www.sciencedirect.com) to give the final results list. Final searches were carried out in July 2013.

Sorting results

We sorted the initial set of results using the title and abstract of each paper. Papers that were not relevant to mangrove fisheries were discarded and the rest were sorted into categories based on the subject matter of the paper. These categories are listed in Table 3, along with a brief description of each one. Data were then extracted from three categories most likely to have data on fishery catches or values: within-mangrove catch data, offshore catch data, and valuations.

Data extraction

Our initial aim was to use the data from existing publications to build a simple model based on actual catch data from fisheries in mangroves. We therefore built a database and populated this with data from the papers in the categories that were most likely to be relevant to this fishery. From each paper, we extracted

- Location data—this included the name, latitude, longitude, and country of the study site (or sites), as well as the size of the study area (the diameter of the smallest circle that could fully enclose

Table 2.—Search terms used following the addition of the third search term set, with the number and relevance of results returned using each set in Web of Knowledge.

Mangrove terms	Fish terms	Action terms	Results	Relevance
mangrove*	cockle*, crab*, finfish, fish*, oyster*, prawn*, shrimp*	aquaculture, capture, catch, farming, fisher*, fishing, gather*, growth, harvest, landing*, nursery, survival, trap*, yield	3,924	64/100
mangrove*	cockle*, crab*, finfish, fish*, oyster*, prawn*, shrimp*	aquaculture, artisanal, capture, catch, farming, fisher*, fishing, gather*, growth, harvest, landing*, nursery, refuge, survival, trap*, valu*, yield	4,044	77/100

Table 3.—The main categories into which results from the literature search were sorted, and the number of results in each category.

Category	Description	Results
Mangrove versus nonmangrove	Comparisons between fish population in mangroves and in adjacent areas.	110
Fish counts	Fish surveys in mangroves or mangrove areas.	450
Within-mangrove catch data	Catch data from fisheries within the mangrove itself.	100
Offshore catch data	Catch data from fisheries offshore of mangrove areas.	10
Links/mechanisms	Papers assessing the links between mangroves and fisheries, and the mechanisms behind these links.	280
Aquaculture	Papers discussing aquaculture in mangrove areas (both yield from aquaculture and the impact of aquaculture on the mangroves).	285
Growth rates of mangrove species	Papers discussing growth rates of fish and invertebrate species found in mangroves, which might be useful for building models.	60
Relevant (other)	Potentially useful papers that do not fit elsewhere.	52
Reviews	Existing reviews of mangrove fisheries literature.	41
Valuations	Papers on the economic valuation of mangrove fisheries, or mangroves more generally.	59

the study site). We divided study sites into four types:

1. Discrete mangrove blocks—the study area includes one discrete block of mangrove bounded by non-mangrove areas.
2. Part of a discrete mangrove block—the study area is within a relatively small, discrete mangrove block, but only includes part of the block.
3. Within a larger mangrove area—the study is set within a large continuous mangrove area (e.g., the Sundarbans or North Brazil).
4. Larger area including multiple mangrove blocks—the study looks at a larger area, which includes multiple mangrove blocks.

Where studies gave a large area as the location, the latitude and longitude of the centroid of this area was stored in the database. Studies for which the location could not be verified were discarded.

- Mangrove species data—Where given, details about the mangrove were recorded, including the tree species present and the community composition as a percentage of each species.
- Catch data—This included the catch method/gear type, the habitat type where the catch was made (e.g., within the mangrove itself, channels within the mangrove, mangrove-fringed lagoon, bay, or estuary) and the start and end date of the study period. We also included all numerical data given about the catch.
- Species data—All the fish and invertebrate species in the catch were recorded. Where proportion or weight of individual species within the catch was given, this was also recorded. We checked fish species against FishBase (www.fishbase.org) and invertebrates against the World Register of Marine Species (www.marinespecies.org), and used currently accepted names and taxonomy where these differed from those given in the paper.

Studies generally reported catches in terms of catch per unit effort, or as catch for the whole fishery for a given time period. Figures reported as catch per unit effort include a value component (physical or monetary), a fishing power component, and a time component, while figures for the entire fishery omit the fishing power component. Where possible, units were standardized as shown in Table 4. Standardization of some units required varying degrees of assumption. Some, such as catches per trap or per fishing trip, could not be standardized and therefore could not be included in the final data sets.

Results

Search results

The final search terms returned a total of 4,358 papers from all three databases once duplicates were removed. Of these 2,766, almost two-thirds of the total set, were discarded as irrelevant. These included papers on mangrove ecology and fish biology, taxonomy and physiology, as well as a significant number of papers totally unrelated to mangrove fisheries that fitted the search terms used by coincidence. The remaining 1592 papers were sorted into categories (Table 3).

Data extraction

In this initial phase, we extracted data from papers in three categories: within-mangrove catch data, offshore catch data, and valuations. These categories were chosen as they contain the studies most likely to have data on fishery catches or values. Of the 169 papers in these three categories, 53 contained data that were included in the database. These 53 studies provided data from 82 study sites in 26 countries. These data can

Table 4.—The main reporting units encountered in the literature for the three dimensions of catch per unit effort, and the standard units that they were converted to. Assumptions required to standardize units are listed as foot notes.

Unit	Standardized unit
Physical/economic units	
g, kg, metric tons, individuals of target species[a]	kg
Currency (11 currencies used in reviewed literature)	2012 US$
Fishing power units	
Fisher, household, village	Fisher
Boat horsepower, boats[b]	Boat horsepower
Trap	Trap
m^2/ha/km^2 fished	ha fished
Time units	
Hour[c], day, month[d], year[e]	Day
Fishing trip	–

[a] Requires knowledge of average weight per individual.
[b] Requires knowledge of average fleet horsepower.
[c] Requires assumption of working hours in each day.
[d] Requires assumption of working days per month.
[e] Requires assumption of working days per year.

be broken down into a variety of different types. Table 5 shows the main data types returned by the review.

The largest data type is catch per unit time, which is largely made up of studies that report the annual catch for an entire fishery within a bay, lagoon, or administrative area. As such, the reporting units are relatively simple, generally giving kilograms per year or individuals per year for some higher-value species such as crabs and mollusks. Value per unit time similarly included mostly annual data on entire fisheries. Catch and value per unit area similarly include whole-fishery statistics, but with the added information of the area that the fishery covers. Studies reporting catch and value for entire fisheries often gave data for multiple years, hence the large discrepancy between the numbers of data in these categories and the numbers of study sites. Catch per unit effort is more complex due to the wide variety of measures of effort (Table 5), many of which cannot easily be compared. The data on density or abundance of fishery species is not directly relevant to this study, as our aim was to gather data on actual catches by commercial fisheries rather than fish abundance.

Fishery types

The data were further divided by the different fishery types that operate in mangroves. If studies reported the species caught in the fishery they were reporting on, we also included this in the database. We used this to categorize fisheries into three main types.

- Inshore mixed species fisheries: These are generally small-scale fisheries, catching a mix of fish, crustaceans, and mollusks. They operate mostly with the mangroves or just offshore. They also tend to be close to settlements as fishers travel on foot or by small boat, often human-powered, limiting the area that they can exploit. Catches are mostly sold in local markets or consumed by the fisher and their family.

Table 5.—The main data types returned by the literature review. For each data type, the table shows the total number of data of that type, the number of study sites with data of that type, the number of countries those sites fall in and the number of studies those data are drawn from. Note that numbers are not additive as many sites have multiple data types.

Data types	Number of data	Study sites	Countries	Studies	Units
Catch per unit area	7	5	4	5	kg/ha/year
Catch per unit effort	67	22	15	20	individuals/fisher/d, g/fisher/d, kg/hour of trawling, kg/1,000 m^2 trawled, metric tons/engine hp/year, kg/net/hour, individuals/trap/d, kg/boat/trip
Catch per unit time	214	43	17	24	kg/year, individuals/year
Value per unit area	27	17	7	9	currency/ha/year
Value per unit effort	5	3	3	3	currency/fisher/d, currency/hour of trawling, currency/net/h, currency/trip
Value per unit time	49	15	13	13	currency/year
Density/abundance/biomass of target species	63	17	6	7	Individuals/m^2, individuals/ha, burrows/100 m^2, g/m^2
Net profit from fishery	26	10	8	10	Currency/year

- Inshore crab and mollusk species: These fisheries are also generally small scale but are more specifically targeted towards single high-value species or species groups. Examples include the mud crab *Scylla serrata* in the Indo-Pacific (e.g., Lebata et al. 2007), the mangrove crab *Ucides cordatus* in South America (e.g., Diele et al. 2005), mangrove cockles *Anadara* spp., and oysters (*Crassostrea* and *Saccostrea* spp.).
- Offshore fisheries: These are larger-scale industrial fisheries operating outside the mangroves and sometimes many miles offshore. This makes them harder to link directly to mangroves, but many of the high-value species that they target are associated with mangroves in their juvenile life history stages. Penaeid prawns have the most well-studied mangrove link (e.g., Rönnbäck et al. 2002; Vance et al. 2002), but other species that use mangroves in clude Barramundi *Lates calcarifer* and various species of snapper (Lutjanidae), mullet (Mugilidae), and sea catfish (Ariidae).

Our aim in this review was to gather data on all of these fishery types. However, because offshore fisheries operate outside the mangroves themselves, fewer papers on this fishery class were identified in our literature search (Table 3) as we only gathered papers with the word "mangrove" in the title, abstract, or keywords. As a result, although we do have some data on these mangrove-linked fisheries, our results are biased towards the fisheries taking place in and around the mangroves themselves.

We are aware that a number of important recreational fisheries also use mangrove areas and would probably represent a fourth distinct category. These can be extremely valuable to national economies; for example, fishing for Bonefish *Albula vulpes*, Permit *Trachinotus falcatus*, and Tarpon *Megalops atlanticus* was worth US$56.5 million to Belize in 2007 (Fedler and Hayes 2008) and $141 million to The Bahamas in 2008 (Fedler 2010). The search terms used in our literature review were geared towards small-scale commercial fisheries, so despite their considerable value, recreational fisheries were not considered in this review.

Preliminary data

Table 6 shows a summary of the values for each of the three fishery types. Inshore mixed species and offshore prawn fisheries both had reasonable numbers of catch data while inshore crab fisheries were less well represented. Reports on inshore mollusk fisheries were available but primarily reported catch per unit effort rather than total catch for the fishery so were not considered further. All three fishery types had a large range of catches, with inshore mixed fisheries having a particularly large range due to a high-yielding fishery in Vietnam (de Graaf and Xuan 1998). Value data were scarcer but showed even more dramatic ranges than the catch data.

The data represent a reasonable geographical spread, especially for inshore mixed species fisheries (Figure 1). Data on crab and prawn fisheries were concentrated in the Indo-Pacific, and there was no data for either of these fisheries from Africa. West Africa is very data-poor in general, with just one data point for a mixed-species fishery in Ghana (Koranteng et al. 2000).

Discussion

Our findings give an indication of the vast body of literature available describing and quantifying the value of mangrove fisheries worldwide, with our initial assessment identifying more than 1,500 published papers relating to mangroves and fisheries. The scientific literature is likely to be a critical source of such information given that many national fishery reporting processes tend to focus on larger-scale commercial fisheries and probably ignore or underreport many mangrove fisheries. Researchers working on the ground in mangrove countries are likely to have the best access to local fisheries reports and statistics, but these data rarely seem to be making the transition from government or nongovernmental organization reports into the scientific literature. As the scientific literature is the most readily available data source to international researchers, reporting local fishery data here would be a key step to allow that data to have an impact through large-scale studies such as this one.

Despite the volume of publications on mangrove fisheries, we found insufficient data to inform the numerical modeling of mangrove fisheries values. Indeed, less than a third of the 169 studies that appeared most promising based on title and abstract actually yielded data on fishery catches. Once these data were collated and divided into data types and fishery types, no category contained more than 16 comparable values (Table 6). The database is likely to yield some further useful information as some of the other categories of paper are reviewed, but we anticipate rapidly diminishing returns from exploring the less relevant categories. The challenge of finding comparable data on mangrove fisheries therefore remains a significant barrier to the development of a model for mangrove fishery values.

A key challenge in drawing together comparable data are the variety of mangrove fishery types. Fisheries vary in target species, scale, and the sector of society involved.

Table 6.—Summary of catch and value data extracted from the literature by fishery type. For each fishery class, the numbers of data on both fishery catch and fishery value are shown, along with the median and range of those catch and value data. Inshore mollusc fisheries were omitted due to insufficient data.

	Catch (kg/ha/year)			Value (US$/ha/year)		
Fishery type	*N*	Median	Range	*N*	Median	Range
Inshore mixed species	16	120.1	20.9–1,195	13	106.1	0.2–2,164.1
Inshore crab	5	45.1	4.1–232.8	1	–	423.4
Offshore prawns	12	122.3	9.3–180.1	2	–	24.3–1,394

This makes modeling challenging, as different fishery types will have different drivers of catch volume and value. For example, the environmental factors that determine the number of prawns produced by a mangrove forest may be different to those determining the number of crabs. Similarly, the socioeconomic variables that determine the fishing effort in a particular fishery will vary with fishery scale and the sectors of society involved. In small-scale fisheries in Southeast Asia, Coates (2002) found that 63–93% of households living near fisheries were involved in fishing, but only a small proportion of these included a professional fisher. In this context, fishing effort would likely be directly proportional to population density, with the entire catch being consumed by fisher's families and communities. By contrast, large-scale industrial fisheries are carried out entirely by professional fishers and have large boats that enable them to operate far from population centers to target high-value species in remote fishing grounds. Their catch is likely to be exported overseas, with little of it being locally consumed. Fishing effort in these fisheries will bear little relation to population density except on very broad scales, instead being primarily driven by the location of the target species.

We have attempted to overcome this challenge through the use of our three broad fishery classes (inshore mixed species, inshore crab and mollusk, and offshore fisheries), which attempt to group fisheries by both target species and scale. This is an imperfect solution; to group only truly similar fisheries would require a classification with

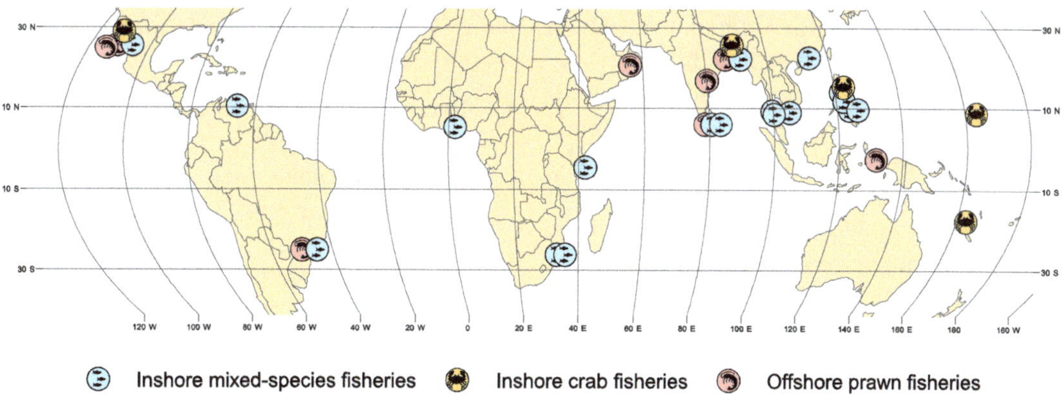

Figure 1.—A map showing the location of data points extracted from the studies returned by the literature review, by fishery type.

multiple dimensions, including boat and gear types and economic, social, and cultural factors (FAO and World Fish Centre 2008), in addition to the target species and scale dimensions that we have used. Each further division of the data, however, reduces the size of the data set that can be used for modeling. Even with our simple three-class system, our largest data set for a single fishery type has just 16 values, making modeling difficult.

The data that we have collated show a large range in both catch and economic value. This is partly the result of the variation in fishery types, even within our categories, but may also genuinely reflect large variation in the catch and value of similar fisheries at different sites. This variability in catch between sites results partly from the mangrove forests themselves; mangroves range from small shrubs on desert margins with low freshwater and nutrient input to 40-m-tall forests in areas with high rainfall and abundant nutrients, with a corresponding variation in the benefits that they provide to fisheries. A further large part of the variation is due to socioeconomic factors. A mangrove in a remote location that is fished by a small village will have a lower catch than one closer to a large city, due to the lower fishing effort. High population density will also give high demand for fish and easy access to markets, driving up the value of the catch. This is demonstrated by the fishery in Mai Po marshes in Hong Kong, which was valued at over $2,000/ha (Qin et al. 2000).

The offshore fishery class poses an additional challenge. Because the fishing grounds are outside the mangroves (in some cases by tens or hundreds of kilometers), it is difficult to directly attribute the catch of those fisheries to the mangroves themselves. Species that have strong evidence of dependence on mangroves include penaeid prawns (Manson et al. 2005) and several snapper species (Kimirei et al. 2013), but even these species use other estuarine and intertidal habitats as well as mangroves. It would be possible to calculate the enhancement that mangroves nursery grounds provide to fisheries by comparing the density of juveniles in mangrove areas with that in areas without mangroves. This analysis has been carried out for oyster reefs (Peterson et al. 2003) and for sea grasses (Blandon and zu Ermgassen 2014a, 2014b), but our literature review did not return sufficient data to do this for mangroves.

A final challenge comes from the geographical and temporal spread of the data. Geographically, we found a reasonable spread of data for mixed species fisheries, although West Africa and South America were both quite data poor. For the other fishery types, however, there were very few data points in South America and none in Africa. Africa was particularly data poor while South America had a number of studies of mangrove crab fisheries, but these did not report the total fishery catch so could not be used. The temporal spread of the data is also a challenge. Fisheries are highly dynamic, with year-to-year variation due to environmental factors as well as human impacts such as overfishing. Data from a 10-year-old study may therefore bear little relation to the state of a fishery today. This should be taken into consideration if these or similar data are used in future analyses.

From our evaluation of the literature, we can make a number of recommendations to scientists working on mangrove fisheries, which would assist in future efforts to review mangrove fisheries values at a large scale:

- Report total catch in addition to catch per unit effort: Catch per unit effort (CPUE) is a key metric in fisheries ecology, but its primary use is in stock assessment. For understanding the value of a particular fishery or the habitat that supports it, the total catch from the fish ery is more useful. The ideal case would be to report both or to report at least catch per unit effort and the total effort for the fishery, from which total catch can be calculated. The area of habitat from which the catch is taken is also helpful.

- Use standard units where possible: A few studies reported catches in individuals or dozens rather than in units of weight. Similarly, studies that reported CPUE included effort units such as catch per net, per trap, or per individual. It may be possible for authors to give sufficient information to convert these into more standard units. For example, for catches reported in individuals of a target species, this would just require a mean weight. There have been efforts to develop standardized units for CPUE (e.g., Salthaug and Godø 2001; Maunder and Punt 2004), but these mostly cover large-scale commercial, boat-based fisheries where boat power can serve as a unit of effort. Units from smaller-scale fisheries are more difficult to standardize, but providing information on the average time spent fishing, the mean number of traps or nets per fisherman, and the number of fishers working in an area would move some way towards this.
- Accurate location is crucial: We discarded a number of studies because they did not give sufficient detail on the study location. Ideally, location should be illustrated on a map showing the sampling points and the limits of the study area. As a minimum, latitude and longitude for the center of the study area should be given. It is also helpful if papers give the size of the study area and the total area of mangrove within it.
- Provide summary statistics for time series: Many papers give data on their studied fishery over a number of years. To use this data in a synthesis, it is necessary to condense the time series to a single value and the best way of doing this depends on the nature of the variability. For example, if the variation is due to environmental stochasticity, then the mean value over the entire study period is likely to give the best summary, whereas if there is a steady declining trend, it would be better to use the most recent value. If the study author has insights into the drivers and direction of any observed trend in catch, it is useful if these are expressed, such that the interpretation of summary statistics is not left to those less familiar with the site.
- Include data from local sources: In reading a number of papers, it was apparent that the authors had used local fishery data, most often from regional or national government bodies but also sometimes from nongovernmental organization reports. This data were not always referenced in a way that allowed us to find it, and even when it was correctly referenced, it was often not available online, meaning that we were unable to access it. Researchers working on the ground with access to these local data sources can dramatically enhance the value of those data and provide an important service to the international research community by, at minimum, reporting the key summary statistics of their data in their research papers.

Next steps

The importance of mangrove to fisheries is already well recognized. To move beyond this recognition towards practical implementation of management measures that might safeguard or even enhance such values, we need a more quantitative understanding of how and why mangrove fisheries values vary. This knowledge can be derived from primary research at an individual site or at a larger scale by a thorough meta-analysis. It is the latter approach we seek to support through summarizing data on fisheries values from the literature.

This work has already proved valuable in helping to determine some of the key drivers of mangrove fisheries, which we used to develop a simple qualitative model for mangrove fishery value (Hutchison et al. 2015). In the future, we hope to move towards a quantitative model by parameterizing our existing model using values from the literature. To do this, we will need to overcome the challenge of the variation in data types. Further research may yield greater insight into potential conversions of our existing data into standard units, which can then be used in further analysis. There is also scope to increase the data set by including gray literature sources and government fisheries reports. Overall, however, this work has shown that even with a comprehensive systematic literature search, data are often not present in ways that allow their use in larger scale analyses. As a result, building a global data set of comparable data on mangrove fisheries remains a challenge.

References

Blandon, A., and P. S. E. zu Ermgassen. 2014b. Corrigendum to "Quantitative estimate of commercial fish enhancement by seagrass habitat in southern Australia" [Estuar. Coast. Shelf Sci. 141 (20 March 2014) 1–8]. Estuarine Coastal and Shelf Science 151:370.

Blandon, A., and P. S. E. zu Ermgassen. 2014a. Quantitative estimate of commercial fish enhancement by seagrass habitat in southern Australia. Estuarine Coastal and Shelf Science 141:1–8.

Coates, D. 2002. Inland capture fishery statistics of Southeast Asia: current status and information needs. Asia-Pacific Fishery Commission, Food and Agriculture Organisation of the United Nations, Regional Office for Asia and the Pacific, RAP publication 11, Bangkok, Thailand.

Collaboration for Environmental Evidence. 2013. Guidelines for systematic review and evidence synthesis in environmental management. Version 4.2. Available: www.environmentalevidence.org/Documents/Guidelines/Guidelines4.2.pdf. (March 2015).

Costello, C., D. Ovando, R. Hilborn, S. D. Gaines, O. Deschenes, and S. E. Lester. 2012. Status and solutions for the world's unassessed fisheries. Science 338:517–520.

de Graaf, G. J., and T. T. Xuan. 1998. Extensive shrimp farming, mangrove clearance and marine fisheries in the southern provinces of Vietnam. Mangroves and Salt Marshes 2(3):159–166.

Diele, K., V. Koch, and U. Saint-Paul. 2005. Population structure, catch composition and CPUE of the artisanally harvested mangrove crab *Ucides cordatus* (Ocypodidae) in the Caeté estuary, North Brazil: indications for overfishing? Aquatic Living Resources 18:169–178.

Ewe, S. M. L., E. E. Gaiser, D. L. Childers, D. Iwaniec, V. H. Rivera-Monroy, and R. R. Twilley. 2006. Spatial and temporal patterns of aboveground net primary productivity (ANPP) along two freshwater-estuarine transects in the Florida Coastal Everglades. Hydrobiologia 569:459–474.

FAO (Food and Agriculture Organization of the United Nations) and World Fish Centre. 2008. Small-scale capture fisheries: a global overview with emphasis on developing countries. A preliminary report of the Big Numbers Project. World Bank, Washington, D.C.

Fedler, A. J., and C. Hayes. 2008. Economic impact of recreational fishing for Bonefish, Permit and Tarpon in Belize for 2007. Friends of Turneffe Atoll, Gainesville, Florida.

Fedler, T. 2010. The economic impact of flats fishing in The Bahamas. The Bahamian Flats Fishing Alliance.

Hutchison, J., D. P. Philipp, J. E. Claussen, O. Aburto-Oropeza, M. Carrasquilla-Henae, G. A. Castellanos-Galindo, M. T. Costa, P. D. Daneshgar, H. J. Hartmann, F. Juanes, M. N. Khan, L. Knowles, E. Knudsen, S. Y. Lee, K. J. Murchie, J. Tiedemann, P. zu Ermgassen, and M. Spalding. 2015. Building an expert-judgment-based model of mangrove fisheries. Pages 17–42 *in* K. J. Murchie and P. D. Daneshgar, editors. Mangroves as fish habitat. American Fisheries Society, Symposium 83, Bethesda, Maryland.

Kimirei, I. A., I. Nagelkerken, Y. D. Mgaya, and C. M. Huijbers. 2013. The mangrove nursery paradigm revisited: otolith stable isotopes support nursery-to-reef movements by Indo-Pacific fishes. PLOS ONE 8(6):e66320.

Koranteng, K. A., P. K. Ofori-Danson, and M. Entsua-Mensah. 2000. Fish and fisheries of the Muni lagoon in Ghana, West Africa. Biodiversity and Conservation 9:487–499.

Laegdsgaard, P., and C. Johnson. 2001. Why do juvenile fish utilize mangrove habitats? Journal of Experimental Marine Biology and Ecology 257:229–253.

Lebata, M. J. H. L., L. L. Vay, J. H. Primavera, M. E. Walton, and J. B. Biñas. 2007. Baseline assessment of fisheries for three species of mud crabs (*Scylla* spp.) in the mangroves of Ibajay, Aklan, Philippines. Bulletin of Marine Science 80:891–904.

Manson, F. J., N. R. Loneragan, G. A. Skilleter, and S. R. Phinn. 2005. An evaluation of the evidence for linkages between mangroves and fisheries: a synthesis of the literature and identification of research directions. Oceanography and Marine Biology: An Annual Review 43:483–513.

Maunder, M. N., and A. E. Punt. 2004. Standardizing catch and effort data: a review of recent approaches. Fisheries Research 70(2–3):141–159.

Peterson, C. H., J. H. Grabowski, and S. P. Powers. 2003. Estimated enhancement of fish production resulting from restoring oyster reef habitat: quantitative valuation. Marine Ecology Progress Series 264:249–264.

Qin, P., Y. S. Wong, and N. F. Y. Tam. 2000. Emergy evaluation of Mai Po mangrove marshes. Ecological Engineering 16:271–280.

Rönnbäck, P., A. Macia, G. Almqvist, L. Schultz, and M. Troell. 2002. Do penaeid shrimps have a preference for mangrove habitats? Distribution pattern analysis on Inhaca Island, Mozambique. Estuarine Coastal and Shelf Science 55:427–436.

Salthaug, A., and O. R. Godø. 2001. Standardization of commercial CPUE. Fisheries Research 49:271–281.

Vance, D., M. Haywood, D. Heales, R. Kenyon, N. Loneragan, and R. Pendreyne. 2002. Distribution of juvenile penaeid prawns in mangrove forests in a tropical Australian estuary, with particular reference to Penaeus merguiensis. Marine Ecology Progress Series 228:165–177.

American Fisheries Society Symposium 83:17–42, 2015

Building an Expert-Judgment-Based Model of Mangrove Fisheries

JAMES HUTCHISON*
Department of Zoology, University of Cambridge
Downing Street, Cambridge CB2 3EJ, UK

DAVID P. PHILIPP
Illinois Natural History Survey, University of Illinois
1816 South Oak Street, Champaign, Illinois 61820, USA
and
Department of Natural Resources and Environmental Sciences
University of Illinois
1102 South Goodwin Avenue, MC 047, Urbana, Illinois 61801, USA
and
Fisheries Conservation Foundation
1816 South Oak Street, Champaign, Illinois 61820, USA

JULIE E. CLAUSSEN
Illinois Natural History Survey, University of Illinois
1816 South Oak Street, Champaign, Illinois 61820, USA
and
Department of Natural Resources and Environmental Sciences
University of Illinois
1102 South Goodwin Avenue, MC 047, Urbana, Illinois 61801, USA

OCTAVIO ABURTO-OROPEZA
Center for Marine Biodiversity and Conservation
Scripps Institution of Oceanography
9500 Gilman Drive, La Jolla, California 92093, USA

MAURICIO CARRASQUILLA-HENAO
Department of Biology, University of Victoria
3800 Finnerty Road, Victoria, British Columbia V8W 3N5, Canada

GUSTAVO A. CASTELLANOS-GALINDO
Marine Programme, WWF Colombia
Carrera 35 No. 4A-25, Cali, Colombia
and
Leibniz Center for Tropical Marine Ecology (ZMT)
Fahrenheitstrasse 6, Bremen 28359, Germany
and

* Corresponding author: jtwh3@cam.ac.uk

Grupo de Investigación en Ecología de Estuarios y Manglares
Departmento de Biología, Universidad del Valle
Calle 13 #100-00, Cali 25360, Colombia

MATTHEW T. COSTA
Center for Marine Biodiversity and Conservation
Scripps Institution of Oceanography
9500 Gilman Drive, La Jolla, California 92093, USA

PEDRAM D. DANESHGAR
Marine and Environmental Biology and Policy Program, Monmouth University
400 Cedar Avenue, West Long Branch, New Jersey 07764, USA

HANS J. HARTMANN
LIttoral, ENvironnement et Sociétés (LIENSs), Université de La Rochelle
UMR 7266 CNRS-ULR, 2 rue Olympe de Gouges, La Rochelle Cedex 17000, France

FRANCIS JUANES
Department of Biology, University of Victoria
3800 Finnerty Road, Victoria, British Columbia V8W 3N5, Canada

MUHAMMED NAEEM KHAN
Department of Zoology, University of the Punjab
New Campus, Lahore 54890, Pakistan

LINDY KNOWLES
Bahamas National Trust
Post Office Box N-4105, The Retreat, Village Road, Nassau, The Bahamas

ERIC KNUDSEN
Ecologists Without Borders
606 Columbia Street NW, Suite 106, Olympia, Washington 98501, USA
and
Sustainable Fisheries Foundation
601 Rainier Street, Snohomish, Washington 98290, USA

SHING YIP LEE
Australian Rivers Institute and School of Environment, Griffith University
Gold Coast Campus, Southport, Queensland 4222, Australia

KAREN J. MURCHIE
Department of Biology, College of The Bahamas
Post Office Box F-42766, Freeport, Grand Bahama, The Bahamas

JOHN TIEDEMANN
Department of Biology, Monmouth University
400 Cedar Avenue, West Long Branch, New Jersey 07764, USA

PHILINE ZU ERMGASSEN
Department of Zoology, University of Cambridge
Downing Street, Cambridge CB2 3EJ, UK

Mark Spalding
Department of Zoology, University of Cambridge
Downing Street, Cambridge CB2 3EJ, UK
and
Global Marine Team, The Nature Conservancy
Department of Zoology, University of Cambridge
Downing Street, Cambridge CB2 3EJ, UK

Abstract.—Mangroves are critically important habitats for fisheries, both for their resident fish, crustacean, and mollusk populations and as nursery grounds for the target species of offshore fisheries. However, the spatial variation in the benefits provided by mangroves to fisheries is poorly understood. Based on expert knowledge of mangrove ecology and fisheries biology, we developed a preliminary model of the spatial distribution of benefits to fisheries from mangroves. The preliminary model covers the environmental factors that determine the amount of fish, crustaceans, mollusks, and other fishery target species produced by mangrove areas (termed "potential fish production") and the socioeconomic variables that determine the level of fishing in any given location. The combination of these two outputs gives the predicted catch. Potential fish production is predicted to be highest where there is high freshwater and nutrient input to mangroves, such as in large estuaries. At large seascape scales, total mangrove area is also an important driver. Fishing effort is highest close to human populations, which provide both the fishers and the markets for their catch. The model is qualitative and has not been parameterized with field data and, as such, should only be considered as a first step towards understanding the spatial variation in the benefits that mangroves provide to fisheries.

Introduction

Mangrove forests are critically important fish habitats, both as nursery grounds (Manson et al. 2005; Nagelkerken et al. 2008) and in supporting large populations of resident fish (Castellanos-Galindo et al. 2013). Mangroves enhance fish communities by providing refuge from predation and sustaining high levels of primary productivity (Alongi 2009), which in turn serves as the basis of a complex food web. The importance of mangroves for fish also makes them important for people. In the Gulf of California, for example, mangrove-related species make up 32% of the small-scale fisheries in the region, giving the mangrove fringe a median value for fisheries of US$37,500/ha (Aburto-Oropeza et al. 2008). Paw and Chua (1991) find a similar figure of 32% for Malaysian fisheries and a much higher figure of 72% for the Philippines.

Mangroves provide a range of other valuable ecosystem services in addition to fish production, including coastal protection, carbon storage, and sediment trapping (Lee et al. 2014). Mangroves provide coastal protection by physically slowing the movement of water as it passes through the aerial root systems (Alongi 2008; McIvor et al. 2012a, 2012b). This function also serves to trap sediments that settle out of the water column as the flow rate decreases (Victor et al. 2004), protecting adjacent ecosystems such as coral reefs and sea grasses that are vulnerable to sedimentation. Carbon is stored both in the aboveground biomass of the mangroves and in the sediments, with soil carbon values averaging 1,023 Mg/ha in Indo-Pacific forests (Donato et al. 2011). The sum value of the many ecosystem services mangrove forests provide is therefore potentially high. For example, in Southeast Asia, ecosystem services from mangroves have been estimated to have a mean value greater than $4,000/ha (Brander et al. 2012).

While mangroves are extremely valuable, they are also highly threatened. As much as a third of the world's mangroves have been lost over the past 50 years, largely through conversion for aquaculture or agriculture, and mangrove deforestation rates have remained among the highest of any forest type (Duke et al. 2007; Spalding et al. 2010; Van Lavieren et al. 2012). Mangrove deforestation has had major impacts on fisheries; in the Gulf of Thailand alone, the cost of mangrove loss to fisheries was estimated at $12,000–408,000 per year in the early 1990s (Barbier et al. 2002). The protection and restoration of mangroves is therefore gaining political and economic support (e.g., Cayubit 2014; Vong 2014).

Our understanding of the value of the ecosystem services that mangroves provide is growing (e.g., Barbier 2012; Siikamäki et al. 2012; Pendleton et al. 2012). Nevertheless, the spatial variability in ecosystem service provision remains poorly quantified for most services, including fisheries. A better understanding of this variability is needed both for more robust valuation and to comprehensively incorporate ecosystem services into management practices (Koch et al. 2009). Information on the spatial distribution of ecosystem service provision would also be valuable for decision makers and conservation practitioners seeking to maximize conservation with limited resources.

Expert judgment is one option for developing ecosystem service models in the absence of complete data. Examples of the use of expert judgment include the development of matrix models that estimate the delivery of a suite of different services for each land cover type within a landscape (e.g., Costanza et al. 1997; Burkhard et al. 2012). However, expert judgment can also be used to develop models for delivery of specific services by specific ecosystems (e.g., Lonsdorf et al. 2009; Haines-Young 2011).

Here, we describe the use of expert judgment to develop a conceptual model of mangrove fishery catches. We then used available global data sets to represent the key drivers in this model, allowing us to qualitatively predict mangrove fishery catches. The model was developed prior to, during, and after a small expert workshop, held within The Second International Symposium on Mangroves as Fish Habitat in Mazatlán, Mexico in April 2014. This is an initial attempt at model development based on a short and relatively simplistic expert consultation process. As such, the results should primarily be considered a proof of concept rather than a finished work, although we believe that even at this stage the model offers some useful insights into the drivers of mangrove fisheries and the spatial variability in the relative importance of this ecosystem service around the globe.

Methods

Conceptual framework

We used a four-stage process to model the benefits that humans derive from fish production by mangroves. The four stages are

1. **Potential fish production:** This is the amount of fish, crustaceans, mollusks, and other fishery target species predicted to be produced by a pristine system, free from human influences. It is based purely on the environmental drivers, of which mangrove extent is one.
2. **Fish production:** This is the amount of fish predicted to be produced by the system, once the influence of human activities such as fishing, pollution, and mangrove degradation is taken into account.
3. **Modeled catch:** This is the amount of fish predicted to be caught. It depends on fish production, as well as fishing effort,

which is driven by socioeconomic factors, particularly population density.

4. **Modeled value:** This is the value of the modeled catch. This can be measured as the market value of the fish caught, which depends on the fish species and on various socioeconomic factors, or it can be measured in terms of the protein that the fishery produces, the people it feeds, or the livelihoods it supports.

In this study, we develop a model for the potential fish production. We then develop a model for fishing pressure and use this to calculate the modeled catch. In this first version of the model, we do not consider the difference between potential fish production (from a theoretical pristine system) and fish production (from a system influenced by human activities such as fishing, pollution and degradation). We also did not incorporate value into our model for reasons discussed below.

Workshop methodology

Prior to the workshop, we compiled a list of potential drivers of mangrove fishery value from a literature review (Hutchison et al. 2014a). We convened an expert workshop to discuss, advise, and drive the selection and weighting of these drivers in the model. Initially, the participants were briefed on the approaches used in developing conceptual models of this sort. A simple conceptual model developed by two of us (J. Hutchison and M. Spalding) prior to the workshop was presented together with an explanation of the sort of data sets that might be used to represent the drivers in the model. Each driver was then discussed to elucidate its relationship and relative importance to mangrove fisheries and thus its role in our overall model. This process had three stages:

1. Discussion of environmental and human impact drivers: This was carried out in three groups, with participants assigned to groups based on their expertise. Following the group discussion, each group described their conclusions to the rest of the participants giving them the opportunity to add further input.
2. Discussion of socioeconomic drivers: This was carried out with all participants present.
3. Voting: The relative importance of each driver in the model was assessed through a voting process. Each participant was given three votes to assign to environmental drivers and two for socioeconomic drivers, reflecting the greater number of environmental variables in the model.

The final selection of drivers for the model was driven by feedback from the workshop on which drivers were most important. However, it was also limited by the drivers for which global spatial data sets were either readily available or could be easily developed from existing data sets.

Environmental drivers.—In our conceptual model (Figure 1), potential fish production is determined by environmental factors. All the environmental drivers considered at the workshop are listed below, along with a summary of the conclusions reached. The voting scores for these drivers are shown in Table 1. Based on the workshop conclusions and voting, we used mangrove area, mangrove biomass (as a proxy for primary productivity), freshwater input, and nutrient input as our driver variables in the model of potential fish production.

- **Mangrove area/linear extent:** The benefits to fisheries from mangroves are likely to be proportional to the area of mangrove. Numerous studies have demonstrated relationships between different aspects of mangrove area and fishery yields in different parts of the world (e.g., de Graaf and Xuan 1998; Carrasquilla-Henao et al. 2013). Habitat benefits to fisheries come primarily from the area of mangroves that are in contact

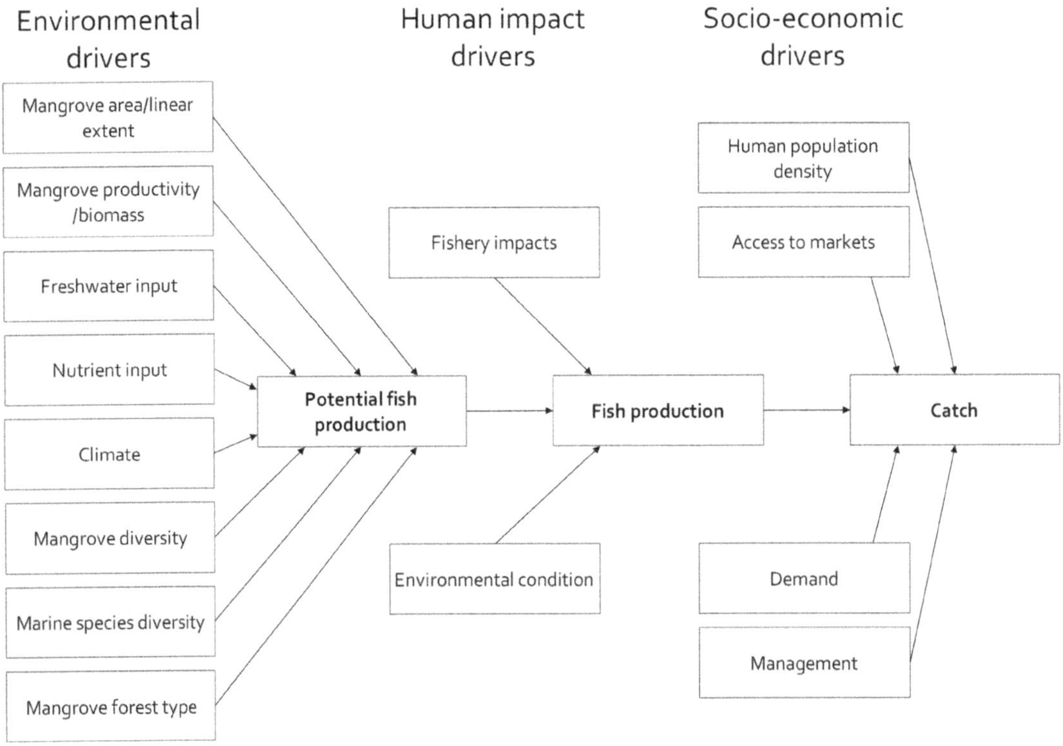

Figure 1.—A conceptual model for the first three stages of mangrove fisheries valuation, showing the drivers that influence each stage. Note that the output of each stage is a driver of the following stage.

with the water, so the fringe area of mangrove may be more important than total area. Recent work in the Gulf of Cal ifornia suggests that there is a constant ratio between fringe area and the square root of the total area (Aburto-Oropeza et al. 2008), but this ratio may be influenced by local geomorphic settings and tidal range (Lee 2004). The consensus among the expert group was that we should use total area rather than fringe area, as even parts of the mangrove that are infrequently inundated may still benefit fisheries through the contribution of leaf litter to the food chain. Because fish production by mangroves is expected to be proportional to mangrove area, we used the other drivers to generate potential fish production and modeled catch per unit area of mangrove, in a 30 arcsecond grid. National and regional estimates were then generated by multiplying these by the area of mangrove in each grid cell, which was derived from the U.S. Geological Survey (USGS) Global Distribution of Mangroves layer (Giri et al. 2011).

- **Primary productivity/biomass:** Primary productivity in mangroves comes from trees, but also from periphyton, microphytobenthos, and phytoplankton. It is likely to be positively correlated with fish production, as it provides the basis for a detrital food chain that ultimately supports species of value to fish eries. Mangrove productivity varies at least 10-fold, from 3.4 Mg ha^{-1} year^{-1} in Florida (Ewe et al. 2006) to 30.9 Mg ha–^{1} year^{-1} in Japan (Khan et al. 2009), due to climatic factors, nutrient availabil-

Table 1.—Environmental drivers: summary of the decisions made by workshop participants and the votes and relative importance (percentage of total votes received) of each driver in the model of potential fish production. Relative importance is the percentage of the total votes received by each driver while the adjusted importance is the percentage of votes received by drivers used in the final model: mangrove primary productivity, nutrient input, and freshwater input. Mangrove extent was also used but only for the purpose of calculating total benefits to fisheries (See Appendix A for detailed methods).

Driver	Relationship	Votes	Relative importance	Adjusted importance
Mangrove extent	Fish catch α mangrove area	12	32.4	-
Mangrove perimeter	Fish catch α mangrove perimeter	2	5.4	-
Mangrove primary productivity	Positive	3	8.1	21.4
Nutrient input	Positive up to a threshold, then plateau, then drop	5	13.5	35.7
Freshwater input	Positive	6	16.2	42.9
Rainfall	Positive but only as part of freshwater input	0	0.0	-
Sea surface temperatures	Possibly upper and lower thresholds, but little influence between these.	0	0.0	-
Mangrove diversity	Possibly positive	1	2.7	-
Fish diversity		3	8.1	-
Mangrove forest type	Probably important, but relationship unclear. Partly accounted for by freshwater and nutrients	5	13.5	-

ity, and forest age. The expert panel felt that primary productivity would positively influence potential fish production but only as a relatively minor driver in the model. There are currently no available maps of mangrove primary productivity, but we used a map of mangrove above-ground biomass as a proxy (Hutchison et al. 2014b).

- **Freshwater input:** Freshwater input has been demonstrated to be positively correlated with fish and prawn production in mangroves and other estuarine habitats (e.g., Vance et al. 1985; Meynecke et al. 2006). The expert panel agreed that freshwater input would positively influence potential fish production, giving it the second greatest number of votes after mangrove area. Maps of sea surface salinity from satellite sensors do exist, but they tend to be unreliable in coastal areas and are generally of poor resolution. Instead, we developed our own map of freshwater influence based on the watersheds and river mouth locations from the USGS HydroSHEDS layer (Lehner et al. 2006) coupled with rainfall data from the World Clim Bioclim data set (Hijmans et al. 2005).
- **Nutrient input:** Nutrient enrichment increases primary production by mangroves and other producers such as phytoplankton and algae, as well as

having some detrimental effects such as increased mangrove mortality (Lovelock et al. 2009). Increased nutrients may also increase the palatability of these producers to grazers (Boyer et al. 2004). Under certain conditions, highly elevated nutrient levels lead to hypoxia and the formation of dead zones (Diaz and Rosenberg 2008), but these are spatially rare and for the purposes of this model they were ignored. The expert panel felt that nutrient input would be positively correlated with fish production by mangroves up to a certain threshold, beyond which production would plateau and then eventually drop. Models that map riverine nutrients have been produced but generally only at very coarse resolution or for a small number of large river mouths. For our model, we therefore used modeled watershed sediment output from the Reefs at Risk project (Burke et al. 2011). This layer was developed to be a proxy of sediment, nutrient, and pollutant delivery, modeled using watershed soil type, slope, land cover, and precipitation.

- **Climate:** Climate is likely to influence fishery productivity through precipitation, which determines freshwater input. Temperature may also be important for some species. In particular, the growth rate of invertebrates such as prawns and crabs are directly affected by water temperature (Staples and Heales 1991). However, the expert panel felt that climate was unlikely to be a major driver of potential fish production except through rainfall, which is already covered as part of freshwater input. Neither rainfall nor sea surface temperature received any votes so climate was not used in the model.
- **Mangrove species diversity:** Diversity of mangrove tree species is higher in the Indo-West Pacific than in the Atlantic East Pacific (Spalding et al. 2010). The expert group felt that mangrove diversity might influence habitat complexity or mangrove productivity, which might in turn influence fish production. However, the level of certainty of this effect was low and mangrove diversity received just one vote so was not used in the model.
- **Marine species diversity:** Like mangroves, diversity of marine species in coastal waters is generally higher in the Indo-West Pacific than in the Atlantic East Pacific (Tittensor et al. 2010). The expert group suggested that a greater diversity of fishery target species may give greater resilience to environmental perturbations and overfishing (Jackson et al. 2001). However, the level of certainty of this effect was low and the relationship between diversity and overall yield was not clear. Global data on marine species diversity do exist (Tittensor et al. 2010) but only at coarse resolutions. For this reason and because of the uncertainty of the relationship between diversity and fish production, marine species diversity was not used in the model.
- **Mangrove forest type:** Lugo and Snedaker (1974) classified mangroves as fringe forest, riverine forest, overwash forest, basin forest, or dwarf forest depending on the ecological setting in which they were found. These different forest types will vary in inundation frequency, carbon dynamics, freshwater, and nutrient availability and may differ in the benefit they provide to fisheries (Flores-Verdugo et al. 2014). Forest type scored highly in the voting process, but consensus in the discussion was that its effect on potential fish production would primarily be through its influence on freshwater and nutrient input. There is also no available data set showing where each forest type is found. For these reasons, forest type was not

used in the model. Freshwater and nutrients are, however, included elsewhere.

Human impact drivers.—These drivers were discussed during the workshop and were felt to be important. However, none of them have global data sets that could be used to represent them in the model. Furthermore, the only metrics we might have used to generate a model for fishery impacts were based around population and markets, which were proposed at the same time as key drivers in determining fishing effort. In the absence of better data or models, we decided to use these data sets only in the fishing effort model. We therefore decided to omit human impacts from this preliminary version of the model. However, a summary of the workshop discussion on each human impact driver is included below.

- **Fishery impacts:** Overfishing and destructive fishing practices will reduce fish populations, which may also impact yield. In many fisheries, years of overfishing have led to greatly reduced catches (Worm et al. 2009). However, catches in many multispecies fisheries such as those found in and around mangroves appear to remain more stable even under elevated fishing pressure, although with a reduction in size and trophic level of the species caught ("fishing down the food chain") (Welcomme 1999). This is reflected in the equation we used to relate modeled catch to fishing effort (Figure A.4), which was developed based on studies in coastal lagoons in West Africa (Laë 1997).
- **Environmental condition:** Mangrove degradation, for example through clearance for timber, changes in hydrological regimes and pollution, will reduce the benefits that the mangroves provide to fisheries (Barbier et al. 2002). Pollution may also influence fishery target species directly by reducing water quality (Alongi 2002) and even driving anoxia in the most heavily polluted estuarine systems (Diaz and Rosenberg 2008).

Socioeconomic drivers.—These drivers were used to model fishing effort. All of the socioeconomic drivers identified were thought to be potentially important by our expert group. However, only human population and markets were used due to a lack of data for the other two proposed drivers. Conclusions of the workshop discussions on these drivers are described below, and the voting scores are shown in Table 2.

- **Human population density:** In general, fishing effort will be positively correlated with human population density as these populations provide both the fishers and the demand for the catch. However, this relationship will vary between different fishery classes. For pure subsistence fishing, there will be a strong relationship with population density in the area immediately surrounding the fishery. However, most small-scale fisheries are still at least partly commercial and therefore require a market for their catch, which requires a population center. Fishing effort is therefore likely to be related to population and distance to the nearest market (Brewer et al. 2009). For the highest value fisheries, there may be little or no relationship to local populations, with fishers traveling long distances to the fishing grounds and exporting their catch to other parts of the country or overseas. Our model is primarily aimed at small-scale fisheries, so we used population as the primary driver of fishing effort, modified by distance to the nearest market. We used population data from the Global Rural-Urban Mapping Project population density grid for our model (CIESIN et al. 2011a). We related fishing effort to catch using an equation developed by Laë (1997) for West African lagoon fisheries.
- **Access to markets:** Most commercial

Table 2.—Socioeconomic drivers: summary of the decisions made at the workshop and the votes and relative importance (percentage of total votes received) of each driver in the model of catch. Only population and access to markets were used in the final model due to a lack of data on management or demand. Note that due to the modelling methods used for catch (see Appendix), the relative importance was not directly used in the model.

Driver	Relationship	Votes	Relative importance
Population	Subsistence fisheries—strongly related to population density. Small- to medium-scale commercial—local population provides fishers but also requires market—town of 10,000 within 2 hours' travel. High value commercial (all scales)—may be unrelated to population.	11	47.8
Access to markets	Will increase catch for small- to medium-scale fisheries.	4	17.4
Management	Positive, by reducing overfishing, but may reduce short-term catches.	3	13.0
Demand	Positive, will increase motivation to fish.	5	21.7

fisheries require a market to sell their catch. In small-scale fisheries, the fishers themselves are most likely to sell to intermediaries, who will then transport the catch to a market in a population center. The expert group reached a consensus that a market was any town of great er than 10,000 people and estimated that fish sellers might transport their catches up to 50 km to reach such a market. This is based on small-scale fisheries with little or no refrigeration. Different assumptions would be required for more industrialized fisheries. Locations of towns of greater than 10,000 people were taken from the Global Rural-Urban Mapping Project settlement points data set (CIESIN et al. 2011b).

- **Demand:** Demand for fish will vary not only with population size, but also with the amount of fish in the diet. This varies by country and across different income groups, influenced by cultural traditions and also by the availability of alternative food sources (Dey et al. 2008). Data on fish consumption are available by country from the Food and Agriculture Organization of the United Nations. However, there is no data at finer at finer spatial scales, so for this reason demand for fish was not included in this version of the model.
- **Management:** Well-managed fisheries can generate higher long-term yields than poorly managed ones (although overfishing may lead to higher short-term yields in poorly managed fisheries). There is very little information available on management measures or their effectiveness, especially for small-scale fisheries, so we did not use management in the model.

Model development

Once the drivers had been filtered based on the workshop discussion and voting, and on the availability of data sets to represent them in the model, the remaining drivers were weighted based on the scores from the workshop voting process. The potential fish production model used mangrove area, freshwater input, nutrient input, and mangrove biomass, while the fishing effort model used human population per unit area of mangrove and whether there was a market within 50 km. Modeled catch was then calculated by multiplying the outputs of these two models. Full details of the model development are given in the Appendix.

Results

We used the models for potential fish production and fishing effort to generate the maps shown in Figure 2. The potential fish production map shows the importance of rivers in our model as sources of freshwater and nutrients, with high potential fish production in areas with many large rivers such as West Africa, but low production in dry areas such as the Arabian Peninsula and Australia. The map of fishing effort reflects the distribution of population density, showing a strong contrast between the sparsely populated north coast of Australia and the high population densities in China, India, and much of Southeast Asia. The modeled catch map is a combination of these two inputs.

We used the maps to generate country level summaries of potential fish production, fishing effort, and modeled catch (Table A.1). The countries with the most important mangrove fisheries are generally those with the largest mangrove areas, including Nigeria, Indonesia, Malaysia, Brazil, and Mexico. Australia and Papua New Guinea are exceptions. Both have large mangrove areas, but low modeled fishing effort due to their low population densities. For the same reason, Indonesia has lower-than-expected modeled catch levels. Despite this, its huge area of mangrove means that it remains one of the top-ranked countries for total catch.

Discussion

Model results

This work represents a first output from an effort to develop a global model and map of the spatial variation in mangrove fisheries. It has important limitations, but we believe it provides a proof of concept for this way of using expert judgment, and that even this current version provides some insights into the importance of mangroves for fisheries.

Not surprisingly, the importance of total mangrove area is highlighted—the six highest ranked countries by total predicted catch from mangrove fisheries all have more than 500,000 ha of mangrove. Fishing effort is also important, as demonstrated by lower catches for Papua New Guinea, Australia, and Indonesia, which all have large areas with low fishing effort. Fishing effort is modeled using population density close to the mangroves, meaning that the highest scoring countries have large mangrove areas that are also close to people and markets. This has important implications for conservation and development; the most valuable areas of mangrove for fisheries are those closest to human populations, which in many places are also likely to be under the greatest threat. Traditionally, conservation effort has emphasized the importance of preserving pristine wilderness areas that remain intact due to their remoteness from people (e.g., Mittermeier et al. 2003; Graham and McClanahan 2013; McCauley et al. 2013). This is in contrast to some of the international agreements where nations have regularly stressed the need to protect not only biodiversity, but also ecosystem services (CBD 2010). To maintain or enhance ecosystem services such as fish production, we need to focus conservation and management efforts particularly on those areas close to people, despite the fact that they are likely to already be far from pristine.

Value and limitations of the model

This first version of the model is conceptual, based on the knowledge and opinions of the authors. It has not been parameterized using field data on fish biomass, fishing effort, or catch, and for this reason it can only be used to make qualitative statements about

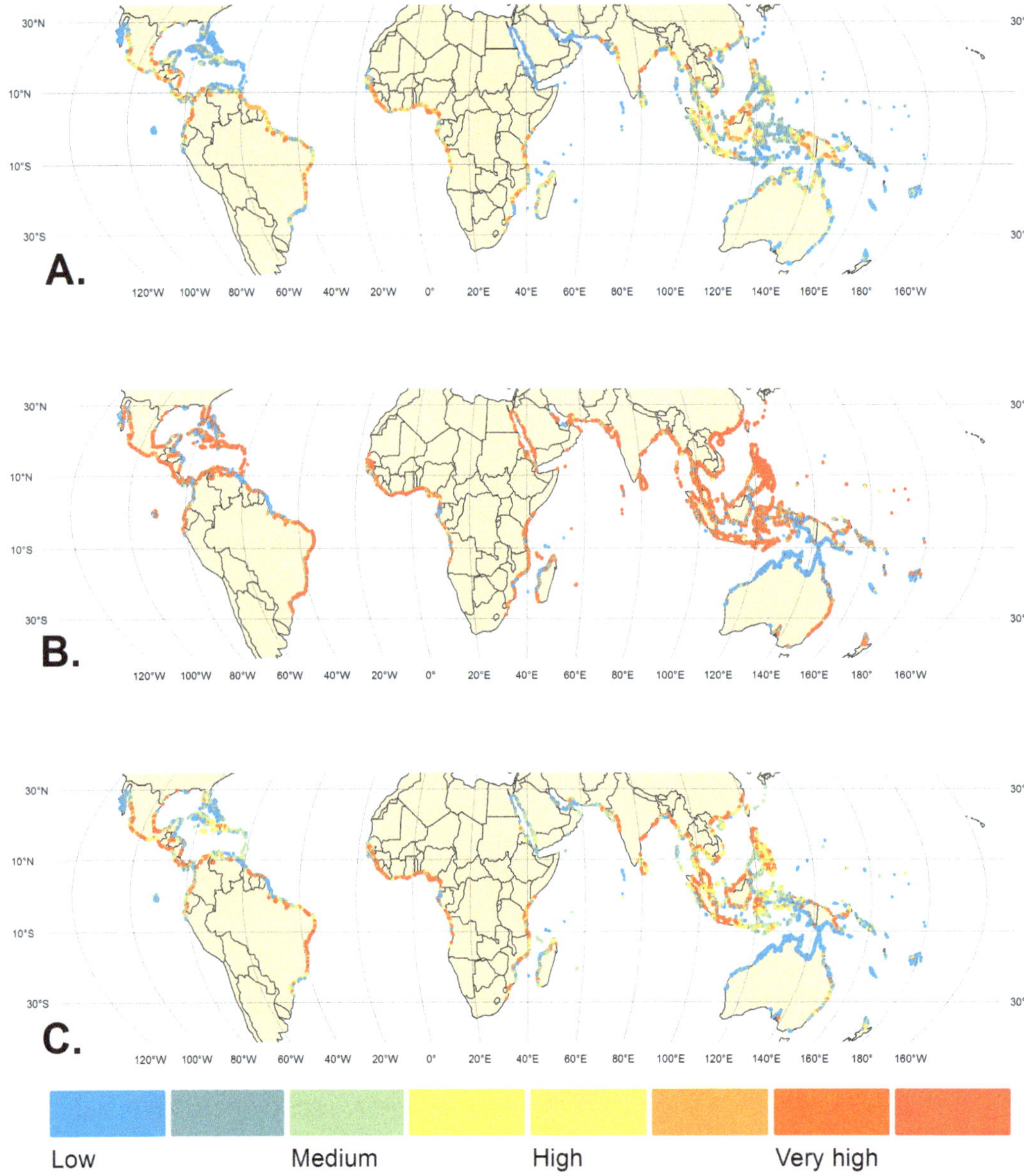

Figure 2.—Maps showing the outputs of the various stages of the model: A. Potential fish production, B. Modeled fishing effort, and C. Modeled catch.

mangrove fisheries. The primary value of the model is as a proof of concept, demonstrating the approach that can be used to elucidate the spatial variation in the ecosystem service of fish production by mangroves. It is a first step to be built upon using more sophisticated modeling techniques and through the use of field data for parameterization.

A key limitation of this first draft model is that it does not attempt to cover the value of mangrove fisheries. The monetary value of a fishery can be expressed as the amount of fish produced multiplied by the price of that fish, minus the cost of its production. However, price and cost are key determinants of effort, which in turn is a major driver of total catch. Accurately representing this circularity requires a more complex modeling approach (e.g., Barbier et al. 2002), but this was both beyond the expertise of our expert panel and beyond the scope of this preliminary modeling attempt. Additionally, value is a complex concept, especially in small-scale fisheries that have a subsistence aspect, and cannot necessarily be fully expressed in monetary terms. Much of the catch may never reach a market or may be traded for other nonmonetary goods. Such values could be expressed in terms of livelihoods supported or dietary protein produced by a fishery.

The elements that we have modeled are based on a number of major simplifying assumptions, a few of which could have a major impact on the model output and are therefore worth highlighting. The first is that the model ignores the negative impacts of people on the ecosystem service through degradation and pollution of the mangroves. This is likely to be important in some locations. Nigeria, for example, is the top-ranked country by total catch predicted by the model. However, much of Nigeria's mangrove area is polluted by oil (Osuji and Ezebuiro 2006), which is likely to heavily impact the fishery. There may also be impacts of overfishing that are not considered in the model; Welcomme (1999) suggests that although yields from small-scale mixed species fisheries remain stable over a range of fishing effort, they can be reduced if fishing effort exceeds a threshold level. Mangrove fisheries also include single species fisheries for crabs and mollusks, and some of these show declining catches due to overfishing (e.g., Beitl 2011). Human impacts are likely to be focused in areas with high population density, which might temper the importance of these areas for fisheries.

The model in its current form is only applicable to small-scale inshore fisheries. For example, it assumes that mangrove fishers will be local residents, which is valid for small-scale fisheries but not for fisheries for high-value species such as penaeid prawns. Northern Australia has very low fishing effort in the model due to its low population density. However, the Northern Prawn Fishery landed 6,600 Mg of prawns along this coastline in 2012 (Barwick 2013), mostly using boats greater than 22 m long, which are able to spend many days at sea and travel many kilometers from their home port to the fishing grounds. A similar modeling approach could be applied to these fisheries but would require the assumptions and data sets for the modeling of fishing effort to be revised. Similarly, the model does not include recreational fisheries, which have very high values in some locations.

Finally, the model only seeks to predict the catch in fisheries in and around mangroves. It does not assess the dependence of those fisheries on the mangrove habitat or the extent to which the presence of mangroves enhances catches. It is true that fisheries would still exist in the absence of mangroves, as other, less-structured estuarine and coastal habitats also produce fish. Indeed, much of the benefit that mangroves provide to fisheries comes through their connections to other habitats such as sea grass beds and coral reefs, with a number of species shown to move between these habitats at different life stages (e.g., Manson et al. 2005; Kimirei et al. 2013). Assessing the extent to which mangroves enhance fisheries over the background level provided by unstructured habitats would require data on fish density inside and outside mangrove habitats, collected using comparable methods (Blandon and zu Ermgassen 2014a, 2014b). We have found very few data meeting these criteria.

Next steps

The model presented in this paper is primarily a proof of concept rather than a finished work and could be improved in a number of ways. A few main areas are obvious targets for further work:

- Improving the conceptual model: This could be achieved through the inclusion of human impacts on mangroves and fish stocks, so that catch can be calculated using modeled fish production rather than potential fish production (see Figure 1). There are also other drivers that may be important but were not included in this version of the model. Fish diversity and mangrove forest type were both identified as potentially important in the workshop, but were not included due to uncertainty over their effects and a lack of suitable data sets to represent them. Tidal range was also mentioned at the workshop but not formally discussed, and has been shown to be correlated with mangrove prawn catch (Lee 2004).
- Improving the use of expert judgment: There is a large body of literature on the best ways of using expert judgment to model ecosystem services. The authors were largely unaware of this literature at the time of the workshop. In future workshops, we would include a measure of confidence in the different modeling steps based on the level of scientific evidence supporting each step, as well as the degree of consensus among the expert group (Jacobs et al. 2015). In addition, the simplistic voting method used may not give an accurate estimate of the weightings attached to each driver in the model. This could be improved using an approach that allows each expert to weight the relative importance of each model driver, ideally as part of a Delphi process allowing for discussion of the initial weightings followed by another weighting opportunity (Martin et al. 2012).
- Developing model versions for different fisheries: in our initial reviews of mangrove fisheries, we identified four broad fishery classes—mixed inshore fisheries, inshore mollusk and crustacean fisheries, offshore fisheries (primarily prawn), and recreational fisheries (Hutchison et al. 2014b). Each of these would have a different set of weightings or even different drivers, particularly for the fishing effort part of the model. The current model focusses primarily on inshore fisheries, a combination of the first two classes. There is an established precedent for using population as a proxy for fishing pressure in small-scale fisheries such as these (e.g., Burke et al. 2011; Teh et al. 2013). For larger-scale fisheries, it is likely to be a much less important factor as the larger boats used are able to travel long distances to remote fishing grounds, and to freeze their catches on board for the return journey.
- Parameterizing the model with fishery catch data: This step would enable us to move from a qualitative model to a numerical model, capable of predicting actual fish catches around the world. We have carried out a literature review and collected data on fish catch and fish abundance in mangroves which might be suitable for this (Hutchison et al. 2014a). However, the data are very variable in quality and the data set is relatively small, with developing countries having particularly poor coverage. We also investigated global databases such as that held by the Sea Around Us Project (Pauly 2007), but we found that the resolution was too coarse for use in a spatial model at the scale of individual mangrove forests. This step therefore represents a significant challenge, but it is an important one: the resulting model could be used to demonstrate the dependence of fisheries in different countries on mangroves and could therefore be a crucial tool to influence policy for mangrove conservation.
- Reporting multiple value metrics: along side efforts to improve and parameterize

the model, it will be important to consider how value can be quantified beyond simple measures of captured biomass. Value can be measured in many ways, and these should include metrics such as livelihoods as well as monetary values.

- Developing fine-scale models for local use: ecosystem services are increasingly used in planning—few countries have detailed knowledge of the value of their mangrove fisheries, but conceptual models could be developed at more local scales, with the advantages of higher resolution data sets and the potential of testing the findings against field data.

The value of mangrove fisheries is widely accepted and yet there are currently no global models or tools to assess such value in specific locations or to reveal global patterns in the variability of this value. At the same time, pressures for development in coastal areas are leading to widespread degradation and loss of mangroves. Better understanding of their value could provide a powerful tool to reduce further mangrove loss and encourage restoration. Through the workshop and subsequent modeling, we are confident that we have developed a valuable proof-of-concept model. With ongoing investment, we hope to build a tool that will generate a model of mangrove fisheries value, but also improve understanding of modeling approaches to enable application at finer scales.

References

Aburto-Oropeza, O., E. Ezcurra, G. Danemann, V. Valdez, J. Murray, and E. Sala. 2008. Mangroves in the Gulf of California increase fishery yields. Proceedings of the National Academy of Sciences of the United States of America 105:10456–10459.

Alongi, D. M. 2002. Present state and future of the world's mangrove forests. Environmental Conservation 29:331–349.

Alongi, D. M. 2008. Mangrove forests: resilience, protection from tsunamis, and responses to global climate change. Estuarine Coastal and Shelf Science 76:1–13.

Alongi, D. M. 2009. The energetics of mangrove forests. Springer Verlag, New York.

Barbier, E. B. 2012. Progress and challenges in valuing coastal and marine ecosystem services. Review of Environmental Economics and Policy 6:1–19.

Barbier, E. B., I. Strand, and S. Sathirathai. 2002. Do open access conditions affect the valuation of an externality? Estimating the welfare effects of mangrove-fishery linkages in Thailand. Environmental and Resource Economics 21:343–365.

Barwick, M. 2013. Northern prawn fishery data summary 2013. NPF Industry Pty, Caloundra, Queensland, Australia.

Beitl, C. M. 2011. Cockles in custody: the role of common property arrangements in the ecological sustainability of mangrove fisheries on the Ecuadorian coast. International Journal of the Commons 5:485–512.

Blandon, A., and P. S. E. zu Ermgassen. 2014a. Corrigendum to "Quantitative estimate of commercial fish enhancement by seagrass habitat in southern Australia" [Estuar. Coast. Shelf Sci. 141 (20 March 2014) 1–8]. Estuarine Coastal and Shelf Science 151:370.

Blandon, A., and P. S. E. zu Ermgassen. 2014b. Quantitative estimate of commercial fish enhancement by seagrass habitat in southern Australia. Estuarine Coastal and Shelf Science 141:1–8.

Boyer, K. E., P. Fong, A. R. Armitage, and R. A. Cohen. 2004. Elevated nutrient content of tropical macroalgae increases rates of herbivory in coral, seagrass, and mangrove habitats. Coral Reefs 23:530–538.

Brander, L., A. Wagtendonk, S. Hussain, A. McVittie, P. H. Verburg, R. S. de Groot, and S. van der Ploeg. 2012. Ecosystem service values for mangroves in Southeast Asia: a meta-analysis and value transfer application. Ecosystem Services 1:62–69.

Brewer, T. D., J. E. Cinner, A. Green, and J. M. Pandolfi. 2009. Thresholds and multiple scale interaction of environment, resource use, and market proximity on reef fishery resources in the Solomon Islands. Biological Conservation 142:1797–1807.

Burke, L., K. Reytar, M. Spalding, and A. Perry. 2011. Reefs at Risk revisited. World Resources Institute, Washington, D.C.

Burkhard, B., F. Kroll, S. Nedkov, and F. Müller. 2012. Mapping ecosystem service supply, demand and budgets. Ecological Indicators 21:17–29.

Carrasquilla-Henao, M., H. A. González Ocampo, A. Luna González, and G. Rodríguez Quiroz. 2013. Mangrove forest and artisanal fishery in the southern part of the Gulf of California, Mexico. Ocean & Coastal Management 83:75–80.

Castellanos-Galindo, G. A., U. Krumme, E. A. Rubio, and U. Saint-Paul. 2013. Spatial variability of mangrove fish assemblage composition in the tropical eastern Pacific Ocean. Reviews in Fish Biology and Fisheries 23:69–86.

Cayubit, R. A. 2014. DENR-8 starts P1-B mangrove rehab project. Available: www.mb.com.ph/denr-8-starts-p1-b-mangrove-rehab-project/. (January 2015).

CBD (Convention on Biological Diversity). 2010. Cop 10 - Tenth meeting of the Conference of the Parties to the Convention on Biological Diversity. Decision X/2. Strategic plan for biodiversity 2011–2020. United Nations Environment Programme, Nagoya, Japan.

CIESIN (Center for International Earth Science Information Network), Columbia University, International Food Policy Research Institute, The World Bank, and Centro Internacional de Agricultura Tropical. 2011a. Global Rural-Urban Mapping Project, (GRUMP), version 1: population density grid. NASA Socioeconomic Data and Applications Center, Palisades, New York.

CIESIN (Center for International Earth Science Information Network), Columbia University, International Food Policy Research Institute, The World Bank, and Centro Internacional de Agricultura Tropical. 2011b. Global Rural-Urban Mapping Project (GRUMP), version 1: settlement points. NASA Socioeconomic Data and Applications Center, Palisades, New York.

Costanza, R., R. d' Arge, R. de Groot, S. Farber, M. Grasso, B. Hannon, K. Limburg, S. Naeem, R. V. O'Neill, J. Paruelo, R. G. Raskin, P. Sutton, and M. van den Belt. 1997. The value of the world's ecosystem services and natural capital. Nature (London) 387:253–260.

de Graaf, G. J., and T. T. Xuan. 1998. Extensive shrimp farming, mangrove clearance and marine fisheries in the southern provinces of Vietnam. Mangroves and Salt Marshes 2:159–166.

Dey, M. M., Y. T. Garcia, K. Praduman, S. Piumsombun, M. S. Haque, L. Li, A. Radam, A. Senaratne, N. T. Khiem, and S. Koeshendrajana. 2008. Demand for fish in Asia: a cross-country analysis. Australian Journal of Agricultural and Resource Economics 52:321–338.

Diaz, R. J., and R. Rosenberg. 2008. Spreading dead zones and consequences for marine ecosystems. Science 321:926–929.

Donato, D. C., J. B. Kauffman, D. Murdiyarso, S. Kurnianto, M. Stidham, and M. Kanninen. 2011. Mangroves among the most carbon-rich forests in the tropics. Nature Geoscience 4:293–297.

Duke, N. C., J.-O. Meynecke, S. Dittmann, A. M. Ellison, K. Anger, U. Berger, S. Cannicci, K. Diele, K. C. Ewel, C. D. Field, N. Koedam, S. Y. Lee, C. Marchand, I. Nordhaus, and F. Dahdouh-Guebas. 2007. A world without mangroves? Science 317:41–42.

Ewe, S. M. L., E. E. Gaiser, D. L. Childers, D. Iwaniec, V. H. Rivera-Monroy, and R. R. Twilley. 2006. Spatial and temporal patterns of aboveground net primary productivity (ANPP) along two freshwater-estuarine transects in the Florida Coastal Everglades. Hydrobiologia 569:459–474.

Flores-Verdugo, F., F. Amezcua, J. M. Kovacs, D. Serrano, and M. Blanco-Correa. 2014. Changes in the hydrological regime of coastal lagoons affect mangroves and small scale fisheries: the case of the mangrove-estuarine complex of Marismas Nacionales (Pacific coast of Mexico). Pages 81–91 *in* F. Amezcua and B. Bellgraph, editors. Fisheries management of Mexican and Central American estuaries. Springer, Dordrecht, Netherlands.

Giri, C., E. Ochieng, L. L. Tieszen, Z. Zhu, A. Singh, T. Loveland, J. Masek, and N. Duke. 2011. Status and distribution of mangrove forests of the world using earth observation satellite data. Global Ecology and Biogeography 20:154–159.

Graham, N. A. J., and T. R. McClanahan. 2013. The last call for marine wilderness? BioScience 63:397–402.

Haines-Young, R. 2011. Exploring ecosystem service issues across diverse knowledge domains using Bayesian belief networks. Progress in Physical Geography 35:681–699.

Hijmans, R. J., S. E. Cameron, J. L. Parra, P. G. Jones, and A. Jarvis. 2005. Very high resolution interpolated climate surfaces for global land areas. International Journal of Climatology 25:1965–1978.

Hutchison, J., A. Manica, R. Swetnam, A. Balmford, and M. Spalding. 2014b. Predicting global patterns in mangrove forest biomass. Conservation Letters 7:233–240.

Hutchison, J., M. Spalding, and P. zu Ermgassen. 2014a. The role of mangroves in fisheries enhancement. The Nature Conservancy, Bethesda, Maryland and Wetlands International, Wageningen, Netherlands.

Jackson, J. B. C., M. X. Kirby, W. H. Berger, K. A. Bjorndal, L. W. Botsford, B. J. Bourque, R. H. Bradbury, R. Cooke, J. Erlandson, J. A. Estes, T. P. Hughes, S. Kidwell, C. B. Lange, H. S. Lenihan, J. M. Pandolfi, C. H. Peterson, R. S. Steneck, M. J. Tegner, and R. R. Warner. 2001. Historical overfishing and the recent collapse of coastal ecosystems. Science 293:629–637.

Jacobs, S., B. Burkhard, T. Van Daele, J. Staes, and A. Schneiders. 2015. "The Matrix Reloaded": a review of expert knowledge use for mapping ecosystem services. Ecological Modelling 295:21–30.

Khan, M. N. I., R. Suwa, and A. Hagihara. 2009. Biomass and aboveground net primary production in a subtropical mangrove stand of Kandelia obovata (S., L.) Yong at Manko Wetland, Okinawa, Japan. Wetlands Ecology and Management 17:585–599.

Kimirei, I. A., I. Nagelkerken, Y. D. Mgaya, and C. M. Huijbers. 2013. The mangrove nursery paradigm revisited: otolith stable isotopes support nursery-to-reef movements by Indo-Pacific fishes. PLOS ONE 8:e66320.

Koch, E. W., E. B. Barbier, B. R. Silliman, D. J. Reed, G. M. Perillo, S. D. Hacker, E. F. Granek, J. H. Primavera, N. Muthiga, S. Polasky, B. S. Halpern, C. J. Kennedy, C. V. Kappel, and E. Wolanski. 2009. Non-linearity in ecosystem services: temporal and spatial variability in coastal protection. Frontiers in Ecology and the Environment 7:29–37.

Laë, R. 1997. Does overfishing lead to a decrease in catches and yields? An example of two West African coastal lagoons. Fisheries Management and Ecology 4:149–164.

Lee, S. Y. 2008. Mangrove macrobenthos: Assemblages, services, and linkages. Journal of Sea Research 59:16–29.

Lee, S. Y., J. H. Primavera, F. Dahdouh-Guebas, K. McKee, J. O. Bosire, S. Cannicci, K. Diele, F. Fromard, N. Koedam, C. Marchand, I. Mendelssohn, N. Mukherjee, and S. Record. 2014. Ecological role and services of tropical mangrove ecosystems: a reassessment. Global Ecology and Biogeography 23:726–743.

Lehner, B., K. Verdin, and A. Jarvis. 2006. HydroSHEDS Technical Documentation, version 1.0. World Wildlife Fund, Washington, D.C. Available: http://hydrosheds.cr.usgs.gov/webappcontent/HydroSHEDS_TechDoc_v10.pdf. (January 2015).

Lonsdorf, E., C. Kremen, T. Ricketts, R. Winfree, N. Williams, and S. Greenleaf. 2009. Modelling pollination services across agricultural landscapes. Annals of Botany 103:1589-1600.

Lovelock, C. E., M. C. Ball, K. C. Martin, and I. C. Feller. 2009. Nutrient enrichment increases mortality of mangroves. PLOS ONE 4:e5600.

Lugo, A. E., and S. C. Snedaker. 1974. The ecology of mangroves. Annual Review of Ecology and Systematics 5:39–64.

Manson, F. J., N. R. Loneragan, G. A. Skilleter, and S. R. Phinn. 2005. An evaluation of the evidence for linkages between mangroves and fisheries: a synthesis of the literature and identification of research directions.

Oceanography and Marine Biology: An Annual Review 43:483–513.

Martin, T. G., M. A. Burgman, F. Fidler, P. M. Kuhnert, S. Low-Choy, M. Mcbride, and K. Mengersen. 2012. Eliciting expert knowledge in conservation science. Conservation Biology 26:29–38.

McCauley, D. J., E. A. Power, D. W. Bird, A. McInturff, R. B. Dunbar, W. H. Durham, F. Micheli, and H. S. Young. 2013. Conservation at the edges of the world. Biological Conservation 165:139–145.

McIvor, A. L., I. Möller, T. Spencer, and M. Spalding. 2012a. Reduction of wind and swell waves by mangroves. Natural Coastal Protection Series: Report 1. The Nature Conservancy, Bethesda, Maryland and Wetlands International, Wageningen, Netherlands.

McIvor, A. L., T. Spencer, I. Möller, and M. Spalding. 2012b. Storm surge reduction by mangroves. Natural Coastal Protection Series: Report 2. The Nature Conservancy, Bethesda, Maryland and Wetlands International, Wageningen, Netherlands.

Meynecke, J.-O., S. Y. Lee, N. C. Duke, and J. Warnken. 2006. Effect of rainfall as a component of climate change on estuarine fish production in Queensland, Australia. Estuarine Coastal and Shelf Science 69:491–504.

Mittermeier, R. A., C. G. Mittermeier, T. M. Brooks, J. D. Pilgrim, W. R. Konstant, G. A. B. da Fonseca, and C. Kormos. 2003. Wilderness and biodiversity conservation. Proceedings of the National Academy of Sciences of the United States of America 100:10309–10313.

Nagelkerken, I., S. J. M. Blaber, S. Bouillon, P. Green, M. Haywood, L. G. Kirton, J.-O. Meynecke, J. Pawlik, H. M. Penrose, A. Sasekumar, and P. J. Somerfield. 2008. The habitat function of mangroves for terrestrial and marine fauna: a review. Aquatic Botany 89:155–185.

Osuji, L. C., and P. E. Ezebuiro. 2006. Hydrocarbon contamination of a typical mangrove floor in Niger Delta, Nigeria. International Journal of Environmental Science & Technology 3:313–320.

Pauly, D. 2007. The Sea Around Us Project: documenting and communicating global fisheries impacts on marine ecosystems. AMBIO: A Journal of the Human Environment 36:290–295.

Paw, J. N., and T.-E. Chua. 1991. An assessment of the ecological and economic impact of mangrove conversion in Southeast Asia. Towards an integrated management of tropical coastal resources. ICLARM Conference Proceedings 22:201–212.

Pendleton, L., D. C. Donato, B. C. Murray, S. Crooks, W. A. Jenkins, S. Sifleet, C. Craft, J. W. Fourqurean, J. B. Kauffman, N. Marbà, P. Megonigal, E. Pidgeon, D. Herr, D. Gordon, and A. Baldera. 2012. Estimating global "blue carbon" emissions from conversion and degradation of vegetated coastal ecosystems. PLOS ONE 7:e43542.

Siikamäki, J., J. N. Sanchirico, and S. L. Jardine. 2012. Global economic potential for reducing carbon dioxide emissions from mangrove loss. Proceedings of the National Academy of Sciences of the United States of America 100:14369–14374.

Spalding, M., M. Kainuma, and L. Collins. 2010. World atlas of mangroves. Earthscan, London.

Staples, D. J., and D. S. Heales. 1991. Temperature and salinity optima for growth and survival of juvenile banana prawns Penaeus merguiensis. Journal of Experimental Marine Biology and Ecology 154:251–274.

Teh, L. S. L., L. C. L. Teh, and U. R. Sumaila. 2013. A global estimate of the number of coral reef fishers. PLOS ONE 8:e65397.

Tittensor, D. P., C. Mora, W. Jetz, H. K. Lotze, D. Ricard, E. V. Berghe, and B. Worm. 2010. Global patterns and predictors of marine biodiversity across taxa. Nature (London) 466:1098–1101.

Van Lavieren, H., M. Spalding, D. M. Alongi, M. Kainuma, M. Clüsener-Godt, and Z. Adeel. 2012. Securing the future of mangroves. A policy brief. Available from United Nations University, Hamilton, Ontario.

Vance, D. J., D. J. Staples, and J. D. Kerr. 1985. Factors affecting year-to-year variation in the catch of banana prawns (*Penaeus mer-*

guiensis) in the Gulf of Carpentaria, Australia. Journal du Conseil 42:83–97.

Victor, S., Y. Golbuu, E. Wolanski, and R. H. Richmond. 2004. Fine sediment trapping in two mangrove-fringed estuaries exposed to contrasting land-use intensity, Palau, Micronesia. Wetlands Ecology and Management 12:277–283.

Vong, M. 2014. Vietnam to clean river with $4 mil mangroves project. Available: www.thanhniennews.com/youth-science/vietnam-to-clean-river-with-4-mil-mangroves-project-24608.html. (January 2015).

Welcomme, R. L. 1999. A review of a model for qualitative evaluation of exploitation levels in multi-species fisheries. Fisheries Management and Ecology 6:1–19.

Worm, B., R. Hilborn, J. K. Baum, T. A. Branch, J. S. Collie, C. Costello, M. J. Fogarty, E. A. Fulton, J. A. Hutchings, S. Jennings, O. P. Jensen, H. K. Lotze, P. M. Mace, T. R. McClanahan, C. Minto, S. R. Palumbi, A. M. Parma, D. Ricard, A. A. Rosenberg, R. Watson, and D. Zeller. 2009. Rebuilding global fisheries. Science 325:578–585.

Appendix A: Detailed Methods and Results

Detailed methods

This appendix contains the detailed methods and equations used to develop the maps for this paper. It should be noted that these methods and models have not been parameterised with any field data and that the results are therefore preliminary. Much of the methodology was informed by the expert judgment of the authors, gathered at a workshop at the *2nd International Symposium on Mangroves as Fish Habitat* in Mazatlán, Mexico in April 2014. The remaining steps were developed by JH and MS based on background knowledge of the underlying ecology of mangrove ecosystems. The quantitative outputs are perhaps a useful pointer to some real patterns but are primarily intended to help ascertain how such a model could be further developed, using both a more extensive expert consultation and field data for parameterization, work we intend to undertake in the future.

Potential fish production.—Based on the workshop results and the availability of data sets, potential fish production per unit area of mangrove was modelled using freshwater input, nutrient input, and mangrove biomass. The layers were weighted based on the percentage of votes they received at the workshop (Table 1): freshwater input received six votes, nutrient input received five, and primary productivity (represented by mangrove biomass) received three (14 votes in total). The resulting weightings were 14/6 = 42.9 for freshwater input, 35.7 for nutrient input, and 21.4 for biomass. Each layer was scaled so that its maximum value was equal to its weighting, giving the modelled potential fish production a theoretical maximum of 100. Details of the development of each layer and of the scaling methods are given below.

- Freshwater input: Freshwater input was calculated using watersheds from the USGS Hydrosheds database (Lehner et al. 2006). We used precipitation data from the WorldClim Bioclim data set to calculate the total volume of precipitation for each watershed, and assumed that all of this volume flowed into the sea at the watershed river mouth. We did not account for evaporation or extraction by humans, which is significant in some regions. The level of freshwater influence was modelled using a Gaussian kernel convolution function to give a smooth decline with distance from each river mouth. The kernel function was calibrated using a map to estimate the range over which freshwater influenced mangroves at several sites with which the authors were familiar, and is shown in Figure A.1. The effect of this freshwater on potential fishable biomass was modelled using a logistic curve which was adjusted based on freshwater values in locations where mangroves are known to be freshwater limited or strongly freshwater influenced. This is based on the assumption that freshwater will benefit fisheries, but that the benefit for a given increase in freshwater will decline as total freshwater input rises. The function also gives a slow initial increase, as small amounts of freshwater will have little effect due to dilution in saltwater. The curve was scaled to give values between 0 and 43, which is the adjusted importance value for freshwater based on the workshop voting (Table 1), rounded to integer values. The curve is shown in Figure A.2.

- Nutrient input: We used modelled watershed sediment output from the Reefs at Risk project (Burke et al. 2011) as a proxy for nutrient input. This layer was developed to be a proxy of sediment, nutrient, and pollutant delivery, modelled using watershed soil type, slope, land cover, and precipitation for some 300,000 watersheds worldwide. It gives a high weighting to agricultural land use as a source of nutrients, but does not include any influence from human, livestock, or industrial waste. We assumed that the dispersion of this sediment would be related to river flow so we used the same Gaussian function to produce the nutrient map as we did for freshwater. As with freshwater, the impact of nutrients on potential fishable biomass was modelled using a logistic curve, this time scaled from 0 to 36. We hypothesised that even small amounts of nutrients from rivers could make a substantial difference to mangrove productivity and fish production in nutrient-poor oceanic mangroves, so we ad-

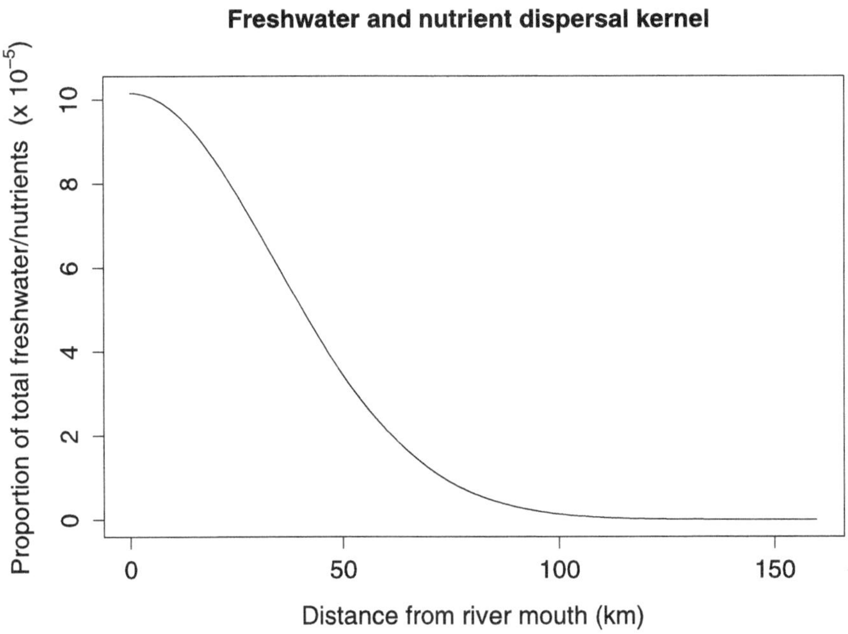

Figure A.1.—The kernel used to generate maps of freshwater and nutrient input from point sources at river mouths. The kernel was roughly calibrated using maps of known riverine influence on mangroves from a small subset of rivers.

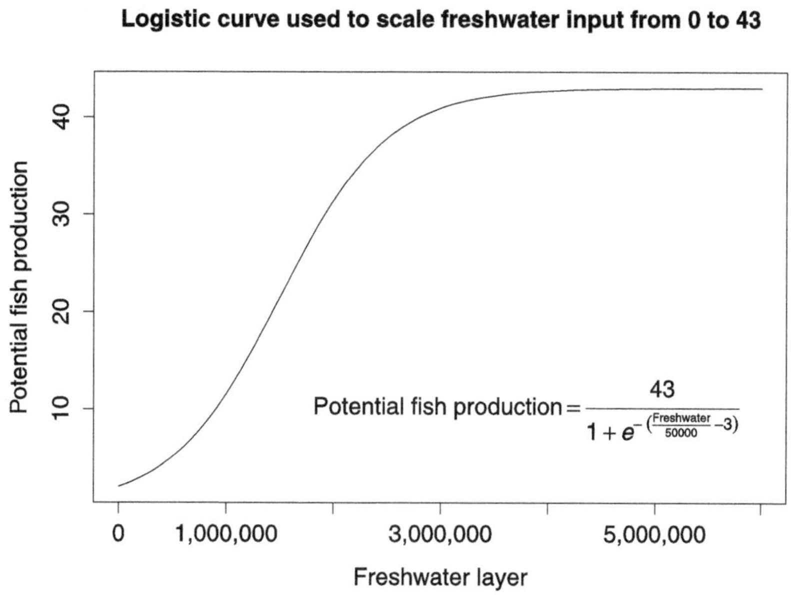

Figure A.2.—The logistic curve used to model the effects of freshwater input on potential fish production. The curve is based on the assumption that freshwater input will increase fish production up to a certain level, at which point it will plateau. The curve is scaled from 0 to 43 to reflect the voting score for freshwater (rounded to integer values).

justed the curve to remove the initial slow increase used for freshwater. Very high nutrient levels would be expected to cause a drop in fish production due to reduced water quality and eutrophication, but this is not included in our model. The curve is shown in Figure A.3.

- Mangrove biomass: Mangrove biomass data were taken from the climate-based model produced by Hutchison et al. (2014). Biomass was assumed to have a linear relationship with fish production. The modelled biomass layer used is in Mg/ha and has a range of 389. It was therefore scaled by multiplying all values by 0.055 to give a maximum value of 21.4, reflecting the weighting of this driver in the model.

The overall equation for potential fish production (PFP) is

$$\mathrm{PFP} = \left(\frac{43}{1 + e^{-\frac{\mathrm{Freshwater}}{50{,}000} - 3}} \right) + \left[-36 + \frac{72}{\left(1 + e^{-\frac{\mathrm{Nutrients}}{7}} \right)} \right] + \frac{\mathrm{Mangrove\ biomass}}{0.055}$$

Fishing effort.—Our fishing effort model used population density adjusted according to whether there was a market within 50 km. Based on consensus in the expert group, markets were defined as towns of 10,000 people or more. The hypothesis underlying the model is that mangrove fisheries without a market within 50 km are likely to be primarily catching fish on a subsistence basis for local consumption, whereas those with access to a market will be catching fish primarily for sale on a commercial basis. We

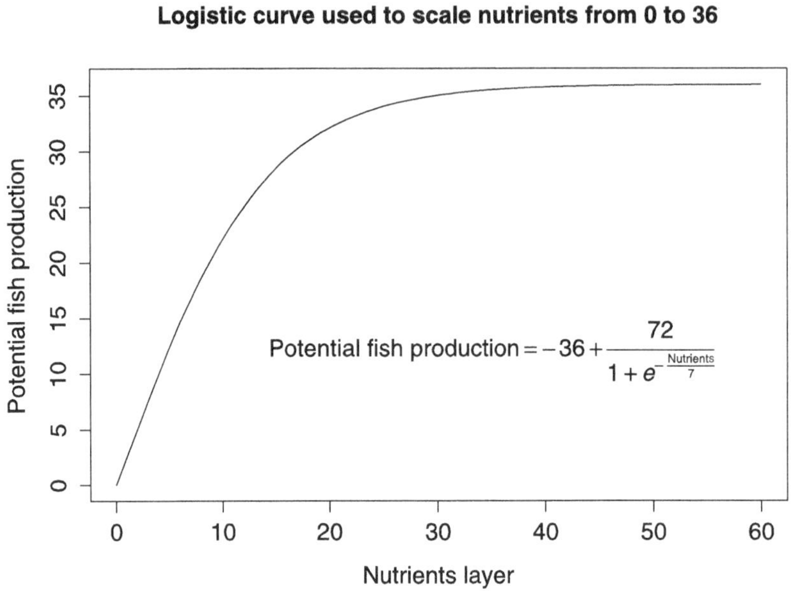

Figure A.3.—The logistic curve used to model the effects of nutrients on potential fish production. As with freshwater, there is assumed to be a positive relationship up to a certain point, followed by a plateau. The curve is scaled from 0 to 36 to reflect the voting score for nutrients (rounded to integer values).

hypothesise that these commercial fisheries will have greater fishing effort per head of population than the subsistence fisheries. Finally, we hypothesise that catch will increase with increasing fishing effort up to a plateau, after which it will remain stable as effort increases further. This relationship is commonly found for small-scale, mixed-species fisheries (Welcomme 1999).

We used the population density grid from the Global Rural-Urban Mapping Project (CIESIN et al. 2011a) layer to calculate the number of people within 10 km of each mangrove grid cell (potential fishers), and the settlement point layer (CIESIN et al. 2011b) to delimit areas within 50 km of a town of 10,000 people or more. Population per unit area of mangrove was calculated by dividing the population within 10 km of each mangrove grid cell by the total area of mangrove within the same 10-km radius. We used a model developed by Laë (1997) from work in coastal lagoon fisheries for the relationship between fishing effort and proportion of the available fish production caught. The model parameters were adjusted to give two separate curves: one for locations within 50 km of a market, and one for locations without a market within 50 km (Figure A.4). The curves represent the proportion of the available fish production caught for a given number of people per unit area of mangrove, where available fish production is some proportion of the modelled potential fish production. This proportion would vary depending on the fisheries biology of the target species but is unimportant in this context as the potential fish production model uses arbitrary units. The parameters for the curves were set based on expert opinion so that 99% of the available fish production was caught when the population within 10 km reached 500 people/km^2 mangrove for areas within reach of a market or 2,000 people/km^2 mangrove without a market. The equations for the two curves are as follows.

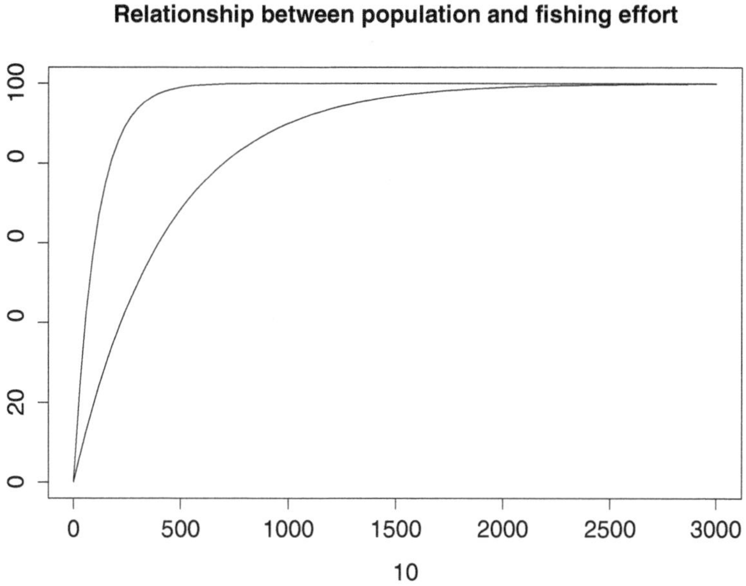

Figure A.4.—The two curves showing our hypothesised relationship between population and the proportion of fish production being caught in locations with and without markets.

With a market,

$$\text{Proportion of fish production caught} = 100\left\{1 - e^{-\left[\frac{-\log(0.01)}{500} \times \text{Population}\right]}\right\}$$

Without a market,

$$\text{Proportion of fish production caught} = 100\left\{1 - e^{-\left[\frac{-\log(0.01)}{2{,}000} \times \text{Population}\right]}\right\}$$

Modelled catch

Modelled catch was calculated by multiplying the potential fish production by the proportion of fish production caught for each mangrove grid cell. As potential fish production and proportion of fish production caught are both scaled from 0 to 100, catch is scaled from 0 to 10,000. We categorized these numbers as very high, high, medium or low based on quartiles to generate the maps shown in Figure 2. National and regional estimates for potential fish production, fishing effort, and catch were generated by multiplying the value of these layers in each grid cell by the area of mangrove in that grid cell, which was derived from USGS Global Distribution of Mangroves layer (Giri et al. 2011). These were then summed by country. Countries were divided into quartiles and scored as low, medium, high, or very high on their mean potential fish production, mean fishing effort, mean catch per unit area of mangrove, and total modelled catch to generate Table A.1.

Appendix References

Burke, L., K. Reytar, M. Spalding, and A. Perry. 2011. Reefs at Risk revisited. World Resources Institute, Washington, D.C.

CIESIN (Center for International Earth Science Information Network), Columbia University, International Food Policy Research Institute, The World Bank, and Centro Internacional de Agricultura Tropical. 2011a. Global Rural-Urban Mapping Project, (GRUMP), version 1: population den-

Table A.1.—Data on mean potential fish production, fishing effort, catch per unit area of mangrove, and total catch, for each country with more than 5,000 ha of mangrove forest. Countries are ordered by total catch and divided into quartiles, which are classed as low, medium, high, and very high.

Country	Mangrove area (ha)	Potential fish production	Fishing effort	Modelled catch	Total modelled catch
Nigeria	744,000	Very high	High	Very high	Very high
Indonesia	2,823,000	High	Low	Medium	Very high
Malaysia	940,000	High	Medium	High	Very high
Brazil	1,057,000	Medium	Medium	Medium	Very high
Mexico	728,000	Medium	Medium	High	Very high
Burma	507,000	High	High	High	Very high
India	393,000	Medium	High	High	Very high
Bangladesh	433,000	Medium	High	High	Very high
Guinea	242,000	Very high	Medium	Very high	Very high
Guinea-Bissau	353,000	High	High	High	Very high
Mozambique	321,000	High	High	High	Very high
Vietnam	217,000	High	Very high	Very high	Very high
Cameroon	158,000	Very high	High	Very high	Very high
Thailand	248,000	Medium	Very high	High	Very high
Sierra Leone	159,000	Very high	High	Very high	Very high
Philippines	258,000	Medium	Very high	Medium	Very high
Colombia	213,000	Very high	Medium	High	Very high
Senegal	168,000	Medium	High	High	High
Cuba	427,000	Low	Medium	Medium	High
The Gambia	70,000	High	Very high	Very high	High
Panama	154,000	High	Medium	Medium	High
Madagascar	272,000	High	Low	Medium	High
Ecuador	137,000	Very high	Medium	High	High
El Salvador	35,000	Very high	Very high	Very high	High
Tanzania	95,000	Very high	Medium	High	High
Australia	962,000	Medium	Low	Low	High
Guatemala	35,000	Very high	Very high	Very high	High
Venezuela	337,000	Medium	Low	Low	High
Honduras	66,000	Very high	Medium	Very high	High
Nicaragua	73,000	High	Medium	High	High
Pakistan	54,000	High	High	High	High
Papua New Guinea	473,000	Very high	Low	Low	High
United States (Continental)	233,000	Low	Low	Low	High
Costa Rica	39,000	Very high	High	Very high	High
Cambodia	47,000	Medium	High	High	Medium
French Guiana	88,000	High	Low	Medium	Medium
Gabon	138,000	High	Low	Low	Medium
Fiji	110,000	Low	Medium	Low	Medium
Brunei	21,000	Very high	Medium	Very high	Medium
Ghana	13,000	Very high	Very high	Very high	Medium
Kenya	33,000	Medium	High	High	Medium

Table A1.—Continued.

Country	Mangrove area (ha)	Potential fish production	Fishing effort	Modelled catch	Total modelled catch
China	18,000	High	Very high	Very high	Medium
Congo, (Kinshasa)	21,000	Very high	Medium	Very high	Medium
Liberia	13,000	Very high	High	Very high	Medium
Belize	57,000	Medium	Low	Medium	Medium
Suriname	75,000	High	Low	Low	Medium
Haiti	15,000	Medium	Very high	Very high	Medium
Angola	27,000	Very high	Low	Medium	Medium
Dominican Republic	18,000	Medium	Very high	High	Medium
Sri Lanka	23,000	Medium	Very high	Medium	Medium
New Zealand	31,000	Low	High	Medium	Medium
Benin	5,000	High	Very high	Very high	Low
Solomon Islands	46,000	Low	Low	Low	Low
Iran	21,000	Medium	Medium	Medium	Low
Guyana	20,000	High	Low	Medium	Low
Jamaica	10,000	Medium	Very high	Medium	Low
Puerto Rico	8,000	Low	Very high	Medium	Low
Trinidad and Tobago	6,000	Low	Very high	Medium	Low
United Arab Emirates	19,000	Low	Very high	Low	Low
Bahamas, The	81,000	Low	Low	Low	Low
Equatorial Guinea	13,000	Low	Low	Low	Low
Cayman Islands	8,000	Low	Very high	Medium	Low
Micronesia, Federated States of	10,000	Low	High	Low	Low
Saudi Arabia	8,000	Low	High	Low	Low
Eritrea	7,000	Low	Medium	Low	Low
Turks and Caicos Islands	17,000	Low	Low	Low	Low
Palau	6,000	Low	Medium	Low	Low
New Caledonia	25,000	Low	Low	Low	Low

sity grid. NASA Socioeconomic Data and Applications Center, Palisades, New York.

CIESIN (Center for International Earth Science Information Network), Columbia University, International Food Policy Research Institute, The World Bank, and Centro Internacional de Agricultura Tropical. 2011b. Global Rural-Urban Mapping Project (GRUMP), version 1: settlement points. NASA Socioeconomic Data and Applications Center, Palisades, New York.

Giri, C., E. Ochieng, L. L. Tieszen, Z. Zhu, A. Singh, T. Loveland, J. Masek, and N. Duke. 2011. Status and distribution of mangrove forests of the world using earth observation satellite data. Global Ecology and Biogeography 20:154–159.

Hutchison, J., A. Manica, R. Swetnam, A. Balmford, and M. Spalding. 2014. Predicting global patterns in mangrove forest biomass. Conservation Letters 7:233–240.

Laë, R. 1997. Does overfishing lead to a decrease in catches and yields? An example of two West African coastal lagoons. Fisheries Management and Ecology 4:149–164.

Lehner, B., K. Verdin, and A. Jarvis. 2006. HydroSHEDS Technical Documentation, version 1.0. World Wildlife Fund, Washington, D.C. Available: http://hydrosheds.cr.usgs.gov/webappcontent/HydroSHEDS_TechDoc_v10.pdf. (January 2015).

Welcomme, R. L. 1999. A review of a model for qualitative evaluation of exploitation levels in multi-species fisheries. Fisheries Management and Ecology 6:1–19.

American Fisheries Society Symposium 83:43–56, 2015

Recreational Fisheries as Conservation Tools for Mangrove Habitats

AARON J. ADAMS*
Department of Biological Sciences, Florida Institute of Technology
Vero Beach Marine Laboratory
805 46th Place East, Vero Beach, Florida 32963, USA
and
Bonefish & Tarpon Trust
Vero Beach Marine Laboratory
805 46th Place East, Vero Beach, Florida 32963, USA

KAREN J. MURCHIE
Department of Biology, College of The Bahamas, Grand Bahama Highway
Post Office Box F-42766, Freeport, Grand Bahama, The Bahamas

Abstract.—Mangrove habitats are among the most threatened coastal habitats. Loss and degradation of mangroves have broad impacts because mangroves are important in prevention of coastal erosion, are excellent at carbon sequestration, and are critical habitats for many marine and estuarine fishes, including coral reef species. However, these species often lack the economic importance or charismatic characteristics to leverage conservation of mangrove habitats. Recent and ongoing research is demonstrating the importance of mangrove habitats to economically important recreational fish species that, in conjunction with reports on the economic impact of these fisheries, provides leverage for mangrove conservation. Two recreational fish species—Common Snook *Centropomus undecimalis* and Tarpon *Megalops atlanticus*—depend on mangrove habitats for one or more of their life stages. Two other species—Bonefish *Albula vulpes* and Barramundi *Lates calcarifer*—are concurrent with mangroves. Each species supports fisheries with annual economic impacts of hundreds of millions of dollars and a large constituency of users. The combination of biological and economic research on these and other recreational fish species can be used as tools for mangrove conservation, adding a much needed boost to coastal conservation efforts and providing protection to many ecologically important but less charismatic species.

Introduction

Mangrove forests are among the most threatened coastal habitats worldwide, decreasing an estimated 35% globally over the past 50 years, with continuing annual declines of 2% (Valiela et al. 2001; Alongi 2002). For example, 16% (11 of 70) of mangrove species examined in an International Union for the Conservation of Nature (IUCN) analysis were classified in one of three categories of threat (critically endangered, endangered, and

* Corresponding author: aaron@bonefishtarpontrust.org

vulnerable; Polidoro et al. 2010). Ongoing and planned coastal development in Belize, Mexico, The Bahamas, and other locations in the Caribbean and worldwide poses serious threats to mangroves that provide essential habitat for many fishes. In fact, the two primary anthropogenic impacts driving fishery collapse are overexploitation and habitat loss or degradation (Lotze et al. 2006).

Economics has been used in recent years as one method to better justify conservation as appropriate, even to people who might not directly interact with the environments in question. For example, the field of environmental economics employs a cost–benefit approach to examining alternative environmental policies, with the goal of identifying the balance between environmental protection for societal benefit and personal benefit driven by market forces (Hanley et al. 2007). To provide a more balanced cost–benefit framework, Costanza et al. (1997) first quantified the economic value of ecosystem services as a means to shift the balance of cost. In general, the commoditization of ecological services has resulted in economically incentivized approaches to increase conservation, such as carbon credits obtained from carbon sequestration (e.g., DFN 2010).

The approach of assigning values to ecological services has generally suffered from two shortcomings. First, large-scale applications do not resonate at a local level where the buy-in of residents is needed for successful habitat conservation. For example, convincing residents of a fishing village that saving their mangrove forest will create carbon credits and reduce global warming is difficult if for no other reason than that it is abstract due to its global scale. Second, there tend to be few economic data available for local or regional resource uses. Indeed, until recent studies by Fedler (2008, 2010, 2011, 2013), there were no economic data available on recreational fisheries in Belize, The Bahamas, and portions of Florida (USA), respectively.

Given the ongoing and increasing threat of habitat loss to coastal ecosystem integrity, there need to be additional tools applied to habitat conservation. There is a strong and growing interest in recreational fishing (Arlinghaus and Cooke 2009), generating important income at regional and national levels (reviewed in Cooke and Suski 2005). With the increasing global interest in recreational fishing, there is risk of an increasing gap between conservation and recreational fishing interests, with potential negative effects on natural resources (Danylchuk and Cooke 2011). However, rather than accept the standard conservation versus fisheries positioning, efforts should be made to engage the recreational fishing community as part of the conservation process (Danylchuk and Cooke 2011 and references therein). This is especially true for species that support catch-and-release fisheries, for multiple reasons: (1) these fisheries generally have low environmental impact because they maintain high fish population size if postrelease fish survival is high (Bartholomew and Bohnsack 2005); (2) the anglers see the connection between habitat health and fishery health—those involved in the fishing industry advocate for habitat conservation; and (3) the per capita economic impact of these fisheries tends to be relatively high, frequently higher than standard tourism (e.g., Fedler 2008, 2010, 2011). Moreover, these fisheries are supported by a large and growing recreational fishing community that is only beginning to engage in conservation, albeit with varying levels of success (Danylchuk and Cooke 2011 and references therein). Furthermore, focusing on catch-and-release fisheries that have a direct connection to threatened habitats may provide the greatest opportunity to reduce the gap between conservation and fisheries and create a platform for a successful collaboration.

This chapter focuses on four marine fish species that support economically important recreational fisheries throughout much of their geographic ranges. Two of these species are obligate users of mangroves and two occur concurrent with mangroves. Through ontogenetic habitat shifts, each of the species uses the wider coastal mangrove habitat mosaic, which makes them appropriate vehicles for multihabitat conservation efforts. The economic importance of these species where recreational fisheries occur provides important leverage for conservation of the habitats that supports the fisheries, leverage that resonates at the local level. Furthermore, their use of the mangrove habitat mosaic qualifies these as umbrella species,[1] and their economic and social importance provide additional local leverage for conservation that is lacking for many other species. In combination, these characteristics support a renewed interest in using sport fish as conservation tools for mangrove habitats.

Focal Species

Tarpon

Tarpon (also known as Atlantic Tarpon) *Megalops atlanticus* is a tropical and subtropical species that is widely distributed in the North Atlantic Ocean, Gulf of Mexico, and Caribbean Sea. Adults are capable of seasonal migrations into warm-temperate regions (e.g., Chesapeake Bay on the eastern coast of North America; Twomey and Byrne 1985), and spawning is presumed to occur offshore (Crabtree et al. 1992; Ault et al. 2008; J. Shenker, Florida Institute of Technology, unpublished data). The planktonic larval duration is 20–40 d, after which time metamorphosing larvae recruit to mangrove wetland habitats.

Tarpon are obligatory users of mangrove habitats because of their juvenile life stage. They use mangrove wetland habitats that are typically low in oxygen (Wade 1962; Dahl 1965; Zerbi et al. 2001; Jud et al. 2011), where they presumably experience low predation and competition for prey (Schlaifer and Breder 1940; Geiger et al. 2000; Seymour et al. 2004). Common nursery characteristics include mangrove or other fringing vegetation that provides structure and protection from bird predators; a mixture of depths—primarily shallow with some deeper pools for fish to congregate when water levels decrease; tidal exchange through narrow, shallow passages that inhibits access by predatory fishes; freshwater inflow; and calm backwaters. Juvenile Tarpon prey upon zooplankton, small crustaceans, polychaetes, and insects that frequent inshore nurseries (Harrington 1958; Jud et al. 2011). As they grow, older juveniles widen their use of habitats to include lagoons, creeks, canals and sloughs, and coastal bays. Adults use a wide range of coastal, estuarine, and riverine habitats (de Menezes and Paiva 1966; Whitehead and Vergara 1978; Chacon-Chaverri 1993; Boujard et al. 1997; Crabtree et al. 1997), which typically include mangrove fringe shorelines. These ontogenetic shifts underscore the importance of a larger mangrove-based habitat mosaic to this species.

Tarpon are economically important throughout their range and support economically valuable recreational fisheries in many locations. For example, the annual economic impact of the recreational Tarpon fishery in the Charlotte Harbor estuary in southwest Florida exceeds US$110 million (Fedler 2011a), and exceeds $19 million in the St. Lucie estuary in southeast Florida (Fedler 2011b). Moreover, Tarpon are an important component of the recreational flats fishery (comprised primarily of Tarpon, Bonefish *Albula vulpes*, and Permit *Trachinotus falcatus*) that has an annual economic impact exceeding $465 million in the Florida Keys (Fedler 2013), $55 million in Belize (Fedler 2013), and $990 million in the Florida Everglades

[1] Umbrella species "are species whose requirements for persistence are believed to encapsulate those of an array of additional species" (Lambeck 1997).

(Fedler 2010b). There are also important recreational fisheries in Cuba, Costa Rica, Nicaragua, Puerto Rico, and other locations that have not been quantified. The economic importance of the recreational Tarpon fishery has been influential in the creation of fishery regulations to protect the fishery: Tarpon are entirely catch and release in Florida (USA), Belize, and Puerto Rico; Tarpon are catch and release only within marine protected areas designated as recreational catch-and-release zones in Cuba; and Tarpon are strictly regulated in Texas and South Carolina (USA).

A recent IUCN scientific assessment classified the Tarpon population as vulnerable (Adams et al. 2013). This classification means that the Tarpon population declined at least 30% in the recent past. This decline occurred for three reasons: harvest primarily in the Caribbean and Gulf of Mexico in the 1960s and 1970s, though recreational harvest within the United States through the 1980s was also a likely contributor; continuing harvest in some areas (e.g., Caribbean coast of Brazil) where a recreational fishery has not been established; and habitat loss and degradation. The IUCN assessment also determined that the decline may continue due to habitat loss and degradation, especially nursery habitats.

Another species of *Megalops*, Oxeye *M. cyprinoides*, may also be worth consideration as a target for use as a conservation tool, but at present, data on fisheries and life history are sparse. The life history of Oxeye, which is widely distributed in the Indo-Pacific Ocean, is similar to Tarpon: spawning occurs offshore (Leis and Reader 1991 as cited in Ley 2008), larval duration is 20–40 d (Tsukamoto and Okiyama 1997; Tzeng et al. 1998), and juveniles occur in estuarine habitats. However, juvenile Oxeye are more abundant in wave-dominated estuaries that are located in higher rainfall catchments and have more constricted mouths, less mangrove area, and sandy substrate (Ley 2005) than Tarpon, so it is unknown whether they are obligate associates of mangroves or facultative.

It is also unclear the geographic extent to which Oxeye would be appropriate as a conservation tool since it supports artisanal fisheries in India and Papua New Guinea (Coates 1987), is commercially targeted in the Philippines and Malaysia, and occurs as bycatch throughout its range. However, there is a recreational fishery in northern Australia (Wells et al. 2003). And since Australia has a good record of recreational angler education about catch-and-release practices and incorporates catch-and-release fisheries into stock assessments and management plans (Sawynok and Pepperell 2002), this may provide a good opportunity to link the fishery and habitat conservation.

Common Snook

The Common Snook *Centropomus undecimalis* is a tropical and subtropical, estuarine-dependent, euryhaline species. Snook are protandrous hermaphrodites (Muller and Taylor 2006) that spawn in aggregations in high-salinity, coastal habitats (Peters et al. 1998; Taylor et al. 1998; Adams et al. 2011), with a planktonic larval duration of approximately 2 weeks (Peters et al. 1998). Recruitment occurs in tidal mangrove wetlands (Peters et al. 1998), and juveniles remain abundant in or near mangrove wetland and creek habitats year-round until age 2, approximately 300 mm (Taylor et al. 2000; Stevens et al. 2007; Barbour and Adams 2012), when they enter the adult population. Adults use estuarine, coastal, and riverine habitats. As with Tarpon, these ontogenetic shifts underscore the importance of a larger mangrove-based habitat mosaic.

Common Snook are obligatory users of mangroves because of their postsettlement and juvenile life stages. However, in regions north of the red mangrove *Rhizophora mangle* geographic range where snook occasionally occur, juvenile Common Snook also use nonmangrove marsh habitats (e.g.,

Juncus maritima). Extensive mark–recapture research showed that 1- and 2-year old juvenile snook show relatively high fidelity to their nursery mangrove creek and that their estimated home range size, and rate of emigration from mangrove creeks, increases with fish size (Barbour and Adams 2012). This expansion of home range and emigration from a nursery creek occurs within a larger habitat mosaic of mangrove creeks, wetlands, ponds, and mangrove fringe shoreline (Barbour and Adams 2012).

Due to their dependence on mangroves, Common Snook populations are threatened by loss of mangrove habitats. However, they are also threatened by anthropogenic alterations of freshwater flows, which can have system-wide impacts (Sklar and Browder 1998; reviewed in Beach 2002). The increase in impervious surfaces and loss of wetland habitats (both due to coastal development) alter the source, timing, and velocity of freshwater flows and thus influence salinity patterns (Sklar and Browder 1998). When more than 10% of a watershed's acreage is converted to impervious surface, the water bodies can become seriously degraded and may be characterized by losses of faunal diversity, productivity, and overall stability (reviewed in Beach 2002), which can have negative impacts on Common Snook. For example, in a study of juvenile Common Snook diet in mangrove creeks in Florida with "highly altered" and "less altered" freshwater flows, the diet diversity and number of snook with prey items in their stomach (versus empty stomachs) were twice as high in the "less altered" creeks, with likely effects on growth and survival (Adams et al. 2009).

In Florida, Common Snook is an economically important recreational species that receives intensive fishing pressure and is under strict management regulation (Taylor et al. 1998). Estimates of the annual economic impact of the recreational Common Snook fishery in Florida vary, but in 2000 the annual economic impact was estimated at $1.56 billion (Taylor et al. 2001). There is also a recreational fishery in Texas, and snook is the focus of recreational and commercial fisheries in Central and South America. In some countries, such as Belize, there is a growing interest in making Common Snook a recreational-only species.

Barramundi

Barramundi *Lates calcarifer*, which is biologically similar to Common Snook, is widely distributed in coastal waters of the Indo-Pacific, from east Africa to southeastern Asia to Australia (as cited in Grey 1987). Like Common Snook, Barramundi spawns in high-salinity coastal waters (as cited in Grey 1987), and they recruit in coastal swamps, wetlands, and other shallow coastal habitats (Russell and Garrett 1988). They utilize wetland habitats and tidal creeks as juveniles (Russell and Garrett 1988) before expanding habitat use as adults to encompass coastal, estuarine, and riverine habitats (Russell and Garrett 1988). As with Tarpon and snook, these ontogenetic shifts underscore the importance of a larger coastal habitat mosaic that includes mangroves.

Also similar to Common Snook, early life stages of Barramundi are impacted by the amount of freshwater flow entering juvenile habitats. In the case of Barramundi, there was a positive correlation between amount of freshwater flow during recruitment and year-class strength (Staunton-Smith et al. 2004). Given the biological similarity with snook, alterations of natural freshwater flow patterns may be problematic for juvenile Barramundi.

The ontogenetic habitat connectivity of Barramundi provides opportunity to link the fishery and habitat conservation. This is most likely in northern Australia, which has an important recreational fishery. Moreover, given the emphasis on increasing catch-and-release survival as part of a national strategy for recreational fisheries management in Australia (Sawynok and Pepperell

2002), the potential link between the fishery and habitat conservation may be especially strong. In fact, in the Northern Territory of Australia, 77% of Barramundi caught in the recreational fishery are released (Henry and Lyle 2003). However, Barramundi is such an important commercial species in Southeast Asia (Tucker et al. 2002; Thirunavukkarasu et al. 2004) that there might not be opportunities to use this species as a conservation tool in this region.

Bonefish

Globally, shallow-dwelling bonefish of the *Albula* genus support popular and highly valuable directed recreational fisheries. The recreational fishery is best developed for Bonefish in the Caribbean, where most data on the fishery are available. Therefore, this discussion will focus on the species of *Albula* that constitutes the Caribbean recreational fishery, Bonefish. However, given the apparent similarity in ontogeny and habitat use across members of the *Albula* genus, and the growing interest for recreational fisheries throughout the genus' range, the information on Bonefish should be applicable to other locations and species, though directed research is needed.

Bonefish is a tropical and subtropical species that inhabits shallow habitats throughout the Caribbean Sea and western North Atlantic (Colton and Alevizon 1983a, 1983b). Ongoing tag–recapture research suggests that adult Bonefish occupy small home ranges (A. J. Adams, unpublished data) and undergo long-distance migrations to spawning locations (Haley 2009; Danylchuk et al. 2011; Adams, unpublished data). Spawning occurs offshore near full and new moons between October and April (Danylchuk et al. 2011). Planktonic larval duration ranges from 42 to 72 d (mean = 53 d; Mojica et al. 1995). Settlement and early juvenile habitats are protected, shallow, open bottoms of sand or sandy mud in bays adjacent to deeper water channels that provide larval access (C. Haak, University of Massachusetts Amherst, personal communication).

Although Bonefish do not appear to be obligatory users of mangrove habitats, they depend upon areas with characteristics that are also required by mangroves: shallow, relatively low-energy marine and estuarine areas protected from direct wave action, such as sand and mud flats, sea grass beds, and tidal creeks (Murchie et al. 2013). Moreover, given their apparent small home range, they are dependent upon intact shallow-water habitat mosaics that include fringing mangrove shorelines and intertidal mangrove flats.

Bonefish supports economically important, catch-and-release, recreational fisheries throughout its range. For example, the annual economic impact of the recreational Bonefish fishery exceeds $141 million in The Bahamas (Fedler 2010a). Bonefish are part of the recreational flats fishery with an annual economic impact exceeding $465 million in the Florida Keys (Fedler 2013) and $55 million in Belize (Fedler 2014). The economic impact of the recreational bonefish fisheries in Cuba, Mexico, Turks and Caicos Islands, and other locations where the recreational fishery is established has not been quantified.

As with Tarpon, the economic value of the recreational fishery has influenced regulators to impose strict regulations to protect the fishery. For example, Bonefish are catch and release only in Florida, Belize, Puerto Rico, and the U.S. Virgin Islands, capture with nets and commercial sale are illegal in The Bahamas, and bonefish are catch and release only within marine protected areas designated as recreational catch-and-release zones in Cuba. Such regulations highlight the importance of these fisheries, which can serve to enhance their use as umbrella species for overall coastal habitat conservation.

Viability of Catch and Release

For recreational fisheries to be effective conservation tools, they must be sustainable, either via tightly regulated harvest (as for

Common Snook in Florida) or as catch-and-release fisheries (as is the case in many areas for Tarpon and Bonefish). For the latter to be effective, postrelease mortality must be low enough that the fishery does not cause a population decline of the target species. However, it is important to realize that the effectiveness of catch and release differs among species, suggesting that species-specific guidelines should be designed for fisheries and that each species is evaluated on the viability of using catch and release as a conservation tool (Cooke and Suski 2005).

Recent research on Tarpon shows that the level of detail needed for guidelines might include fishing location, angler behavior, and fish size. For example, the estimated postrelease mortality of Tarpon on the Gulf of Mexico coast of Florida is 5% (Guindon 2011). However, predation by sharks increases postrelease mortality to 13%, and postrelease mortality can differ among estuarine systems (Guindon 2011) in locations where Tarpon aggregations are targeted by high densities of recreational anglers. In most of these aggregation locations, shark abundance is naturally high, and there is concern that the high fishing effort increases abundance of sharks and rates of fishing-related predation. Outside of these locations, predation by sharks appears to be low, but additional research is warranted. Finally, large Tarpon (>1 m fork length) show physiological stress with handling, whereas Tarpon less than 1 m exhibited little physiological stress due to handling (Guindon 2011).

Common Snook also respond well to catch and release. Overall postrelease mortality was estimated at 2.13% with mortalities attributed to hook placement (40% of snook that died had been hooked in the stomach or throat (Taylor et al. 2001). In addition, there was no mortality of snook during treatments that examined the effects of handling and tagging (Taylor et al. 2001).

Similarly, Barramundi have high postrelease survival rates, but there is seasonal variation in mortality. Although the overall postrelease survival rate was 90.5%, survival was 100% in winter and 80% in summer (de Lestang et al. 2004), which should be taken into consideration in education and management plans.

Research on Bonefish reveals the importance of handling practices, which has direct implications for education and angler behavior. The major factors affecting postrelease survival of Bonefish involve the amount and type of handling by anglers, exposure to air, duration of the fight, and abundance of predators at the release location (Cooke and Philipp 2004; Danylchuk et al. 2007a, 2007b). When Bonefish are improperly handled and exposed to air for extended durations, their mortality increases significantly (Danylchuk et al. 2007a, 2007b). In areas with few predators, the postrelease survival of properly handled fish is greater than 95% (Cooke and Philipp 2004; Danylchuk et al. 2007a) but can drop to 61% if predatory sharks are abundant (Cooke and Philipp 2004).

For each of these species, overall postrelease mortality is low, indicating that catch and release is a viable conservation tool for these fisheries. In addition, caution must be applied to account for the effects of predators and angler behavior. Education of anglers and fishing guides is important to ensure high postrelease survival. For example, education should be used to adjust angler and guide behavior to reduce handling of fish prior to release, minimize air exposure, and avoid locations with high predator abundance or cease fishing and move to other locations if predators appear.

As potential conflicts between conservation and recreational fishing sectors increase, concurrent with an increasing need for natural resource conservation, efforts must be made to engage the recreational fishing sector in conservation efforts (Danylchuk and Cooke 2011). For example, marine-protected areas might be managed as catch-and-release zones that allow recreational catch-and-release fisheries (Cooke et al. 2006) that support local economically important fisheries. In fact, a recent model suggested that catch-and-release zones would be effective spatial management tools for Common Snook, resulting in higher spawning potential ratio, higher abundance of trophy-sized snook, and overall higher abundance (Ley and Allen 2013). If such catch-and-release zones are paired with appropriate habitat conservation, multiple conservation objectives can be obtained. However, if such an approach is to be effective, the effort must also include the dissemination of proper catch-and-release guidelines by management agencies (Pelletier et al. 2007) and others (Sawynok and Pepperell 2002) because education of recreational anglers on best handling practices and on ways to reduce predation will play an important role in the effectiveness of catch-and-release fishing as a management tool (Cooke et al. 2006).

Applications to Conservation

Because Tarpon, Common Snook, Bonefish, and Barramundi rely on mangrove habitats for some portion of their life history, or upon the productivity of mangroves as sources of prey, the loss and degradation of mangrove habitats negatively impacts the fisheries that rely on these species. For Tarpon and Common Snook, whose juveniles require shallow, protected mangrove swamps, habitat loss and degradation causes a decline in the amount of available essential nursery habitat, which has negative implications for adult populations and thus the fisheries. Barramundi appear to have more plasticity in habitat use, but also depend on the habitat mosaic associated with mangroves. Similarly, although Bonefish do not appear to rely on mangrove habitats as juveniles, they do use mangrove habitats for foraging and refuge as adults and appear to have similar abiotic requirements to mangroves. Given the importance of mangroves to Tarpon, Common Snook, Bonefish, Barramundi, and their prey, mangrove loss is likely having major negative consequences on population resilience, which negatively impacts the economic value of these fisheries.

Anthropogenic fragmentation of the coastal habitat mosaic that is anchored by mangroves is also a threat. Habitat fragmentation occurs when portions of a continuous habitat, or of habitat mosaics, are destroyed or degraded, creating gaps in suitable habitat. For example, fragmentation of tidal mangrove creeks of the Bahamas simplified food webs (Layman et al. 2007). This can have population and system-level implications because reductions in prey diversity cause sublethal losses in productivity, including reduced growth rates, lower fish condition, and higher parasite loads of top predators (Rypel and Layman 2008). This is especially important for fishes that use multiple habitats through their life history strategies, as do the species addressed here.

In other words, as go the habitats, so go the fisheries. Indeed, the fish–mangrove habitat association has been demonstrated for many species (e.g., coral reef fishes, Mumby et al. 2004). However, such fish–habitat associations have often been insufficient to prevent loss of mangrove habitats. In contrast, the economic importance of Tarpon, Common Snook, Bonefish, and Barramundi and the cultural importance of the fisheries enable them to be used as tools for habitat conservation. Most important, conservation efforts that affect these species engage a large user group (recreational anglers and fishing guides) that at present is only partially involved (Danylchuk and Cooke 2011). Such users will advocate for habitat conservation that helps to protect their fishery, particularly when the scale of conservation efforts are local with local benefits (Cowx et al. 2010).

Ecotourism is becoming recognized as an important component of sustainable economies, especially in the developing world where Tarpon, Common Snook, bonefishes, and Barramundi occur. Moreover, it is in these same regions where the greatest loss of mangroves is occurring (Polidoro et al. 2010). Recreational fishing, especially catch-and-release fishing, is an important component of ecotourism and can make a significant economic contribution while maintaining high abundance of targeted fishes if managed appropriately. As previously stated, in locations where it is established, the activity makes significant contributions to local economies.

At present, these catch-and-release fisheries tend to occur without government support and, in most cases, without support of conservation organizations. In other areas, such fisheries could be encouraged to develop, with a focus on locations of high priority conservation focus. However, great care must be taken to avoid overcapitalization and overfishing of the resource, which has negatively impacted other fisheries. For example, too much fishing effort might result in a decline in catch rates because fish may become educated to angler activities, thus reducing catchability, causing a decline in the quality of the fishery and a loss of clientele, even if the abundance of the target species does not decline. Species- and location-specific research is needed to identify fisheries capacity to avoid this scenario.

Cowx et al. (2010) posit that recreational angler buy-in is likely if management scales are small, conservation threats originate from external sources, and ecological awareness of the need for conservation is high. The nature of the Tarpon, Snook, Bonefish, and Barramundi fisheries allow for focusing conservation efforts at local (small) scales because the fisheries tend to be location-based. For example, recreational anglers traveling to fish for bonefish target specific destinations, such as the Bahamas. The top threat to the fisheries, and the top perceived threat by anglers, is habitat loss and degradation, which is external to the fishery and is mostly due to coastal development and associated water quality issues. Recreational anglers and guides engaged in catch-and-release fisheries tend to be advanced anglers who understand the connection between the health of the environment and quality of the fishery (Oh and Ditton 2006), so are well positioned to advocate for conservation. Clearly, the factors suggested by Cowx et al. (2010) are applicable for the fisheries addressed herein.

Finally, the benefit from a conservation perspective is that these are umbrella species for conservation of the mangrove-based habitat mosaic, thus providing conservation protections for all species that use these habitats, whether or not they are charismatic. The economic and cultural value of these recreational fisheries, as well as their charismatic characteristics (Bonefish, Tarpon, and Common Snook, for example, regularly appear on fishing magazine covers and are the focus of articles in these magazines), make them uniquely useful to conservation efforts. The benefit from the recreational fisheries perspective is that habitat conservation will greatly contribute to a healthy and sustainable fishery. The concurrence of threatened coastal habitats and economically important recreational fisheries create a good scenario for collaborative fisheries and habitat conservation.

Acknowledgments

This chapter resulted from conversations and field work that occurred over many years of collaborative research. We thank the long list of colleagues that were part of those discussions and projects, as well as those who organized and hosted the *2nd International Symposium on Mangroves as Fish Habitat*. We also thank two anonymous reviewers for comments that improved this manuscript. A. J. Adams was funded by Bonefish & Tarpon Trust.

References

Adams, A. J., R. K. Wolfe, and C. A. Layman. 2009. Preliminary examination of how human-driven freshwater flow alteration affects trophic ecology of juvenile Snook (*Centropomus undecimalis*) in estuarine creeks. Estuaries and Coasts 32:819–828.

Adams, A. J., J. E. Hill, and C. Samoray. 2011. Characteristics of spawning ground fidelity by a diadromous fish: a multi-year perspective. Environmental Biology of Fishes 92:403–411.

Adams, A. J., A. Z. Horodysky, R. S. McBride, T. C. MacDonald, J. Shenker, K. Guindon, H. D. Harwell, R. Ward, and K. Carpenter. 2013. Conservation status and research needs for tarpons (Megalopidae), ladyfishes (Elopidae), and bonefishes (Albulidae). Fish and Fisheries 15:280–311.

Alongi, D. M. 2002. Present state and future of the world's mangrove forests. Environmental Conservation 29:331–349.

Arlinghaus, R., and S. J. Cooke. 2009. Recreational fishing: socioeconomic importance, conservation issues and management challenges. Pages 39–58 *in* B. Dickson, J. Hutton, and B. Adams, editors. Recreational hunting, conservation and rural livelihoods: science and practice. Blackwell Publishing, Oxford, UK.

Ault, J. S., R. Humston, and M. F. Larkin. 2008. Population dynamics and resource ecology of Atlantic Tarpon and Bonefish. Pages 216–258 *in* J. S. Ault editor. Biology and management of the world tarpon and bonefish fisheries. CRC Press, Boca Raton, Florida.

Barbour, A. B., and A. J. Adams. 2012. Biologging to examine multiple life stages of an estuarine-dependent fish, Common Snook, *Centropomus undecimalis*. Marine Ecology Progress Series 457:241–250.

Bartholomew, A., and J. A. Bohnsack. 2005. A review of catch-and-release angling mortality with implications for no-take reserve. Reviews in Fish Biology and Fisheries 15:129–154.

Beach, D. 2002. Coastal sprawl: the effects of urban design on aquatic ecosystems in the United States. Pew Oceans Commission, Arlington, Virginia.

Boujard, T., M. Pascal, F. J. Meunier, and P. Y. Le Bail. 1997. Poissons de Guyane. Guide ecologique de l'approuague et de la reserve des nouragues. [Fish of Guyana. Ecological guide of the Approuague and the Nourages Reserve.] Institut National de la Recherche Agronomique, Paris.

Chacon-Chaverri, D. 1993. Aspectos biometricos de una poblacion de sabalo, Megalops atlanticus (Pisces: Megalopidae). [Biometric aspects of a population of Tarpon, *Megalops atlanticus* (Pisces: Megalopidae).] Revista Biologia Tropical 41:13–18.

Coates, D. (1987) Observations on the biology of the Tarpon, *Megalops cyprinoides* (Broussonet) (Pisces:Megalopidae) in the Sepik River, northern Papua New Guinea. Australian Journal of Marine and Freshwater Research 38:529–535.

Colton, D. E., and W. S. Alevizon. 1983a. Feeding ecology of Bonefish in Bahamian waters. Transactions of the American Fisheries Society 112:178–184.

Colton, D. E., and W. S. Alevizon. 1983b. Movement patterns of the Bonefish (*Albula vulpes*) in Bahamian waters. U.S. National Marine Fisheries Service Fishery Bulletin 81:148–154.

Cooke, S. J., and D. P. Philipp. 2004. Behavior and mortality of caught-and-released Bonefish (*Albula* spp.) in Bahamian waters with implications for a sustainable recreational fishery. Biological Conservation 118:599–607.

Cooke, S. J., and C. D. Suski. 2005. Do we need species-specific guidelines for catch-and release recreational angling to conserve diverse fishery resources? Biodiversity and Conservation 14:1195–1209.

Cooke, S. J., A. J. Danylchuk, S. E. Danylchuk, C. D. Suski, and T. Goldberg. 2006. Is catch-and-release angling compatible with no-take marine protected areas? Ocean and Coastal Management 49:342–354.

Costanza, R., R. d'Arge, R. de Groot, S. Farber, M. Grasso, B. Hannon, K. Limburg, S. Naeem, R. V. O'Neill, J. Paruedo, R. G. Raskin, P. Sutton, and M. van den Belt. 1997. The value of the world's ecosystem services and natural capital. Nature (London) 387:253–260.

Cowx, I.G., R. Arlinghaus, and S. J. Cooke. 2010. Harmonizing recreational fisheries and conservation objectives for aquatic biodiversity in inland waters. Journal of Fish Biology 76:2194–2215.

Crabtree, R. E., E. C. Cyr, R. E. Bishop, L. M. Falkenstein, and J. M. Dean. 1992. Age and growth of Tarpon, *Megalops atlanticus*, larvae in the eastern Gulf of Mexico, with notes on relative abundance and probable spawning areas. Environmental Biology of Fishes 35:361–370.

Crabtree, R. E., D. C. Chaverri, W. O. McLarney, and J. M. Dean. 1997. Reproduction of Tarpon, *Megalops atlanticus* from Florida and Costa Rican waters and notes on their age and growth. Bulletin of Marine Science 61:271–285.

Dahl, G. 1965. La metamorfosis desde leptocephalus hasta estado postlarval en el sabalo *Tarpon atlanticus* (V. Cuv et al.). [The metamorphosis from leptocephalus to post-larval stage in Tarpon, T*arpon atlanticus* (V. Cuv et al.).] Autonoma Reginal de los Valles del Magdalena y Sin u. 1–20.

Danylchuk, A. J., and S. J. Cooke. 2011. Engaging the recreational angling community to implement and manage aquatic protected areas. Conservation Biology 25:458–464.

Danylchuk, A. J., S. J. Cooke, T. L. Goldberg, C. D. Suski, K. J. Murchie, S. E. Danylchuk, A. Shultz, C. R. Haak, E. Brooks, A. Oronti, J. B. Koppelman, and D. P. Philipp. 2011. Aggregations and offshore movements as indicators of spawning activity of Bonefish (*Albula vulpes*) in the Bahamas. Marine Biology 158:1981–1999.

Danylchuk, A. J., S. E. Danylchuk, S. J. Cooke, T. L. Goldberg, J. B. Koppelman, and D. P. Philipp. 2007a. Post-release mortality of Bonefish (*Albula vulpes*) exposed to different handling practices during catch-and-release angling in South Eleuthera, Bahamas. Fisheries Management and Ecology 14:149–154.

Danylchuk, S. E., A. J. Danylchuk, S. J. Cooke, T. L. Goldberg, J. Koppelman, and D. P. Philipp. 2007b. Effects of recreational angling on the post-release behavior and predation of Bonefish (*Albula vulpes*): the role of equilibrium status at the time of release. Journal of Experimental Marine Biology and Ecology 346:127–133.

de Lestang, P., R. K. Griffin, and Q. A. Allsop. 2004. Assessment of the post-release survival and stress physiology of Barramundi (*Lates calcarifer*). Northern Territory Government, Fishery Report No. 73, Darwin, Northern Territory, Australia.

DFN (Danone Fund for Nature). 2010. Achieving carbon offsets through mangroves and other wetlands. November 2009 expert workshop meeting report. Nick Davidson, editor. Danone Group/IUCN/ Ramsar Convention Secretariat, Gland, Switzerland.

Fedler, A. 2010a. The economic impact of flats fishing in the Bahamas. Report to the Everglades Foundation, Palmetto Bay, Florida.

Fedler, A. 2010b. The economic impact of recreational fishing in the Everglades region. Report to the Everglades Foundation, Palmetto Bay, Florida.

Fedler, A. 2011a. The economic impact of recreational Tarpon fishing in the Caloosahatchee River and Charlotte Harbor region of Florida. Report to the Everglades Foundation, Palmetto Bay, Florida.

Fedler, A. 2011b. The economic impact of recreational Tarpon fishing in the St. Lucie River and Treasure Coast region of Florida. Report to the Everglades Foundation, Palmetto Bay, Florida.

Fedler, A. 2013. Economic impact of the Florida Keys flats fishery. Report to Bonefish & Tarpon Trust, Key Largo, Florida.

Fedler, A. 2014. Economic impact of flats fishing in Belize. Report to Bonefish & Tarpon Trust, Key Largo, Florida.

Geiger, S. P., J. J. Torres, and R. E. Crabtree. 2000. Airbreathing and gill ventilation frequencies in juvenile Tarpon, *Megalops atlanticus*: responses to changes in dissolved oxygen, temperature, hydrogen sulfide, and pH. Environmental Biology of Fishes 59:181–190.

Grey, D. L. 1987. An overview of *Lates calcarifer* in Australia and Asia. Pages 15–21 *in* J. J. Copland and D. L. Grey, editors. Management of wild and cultured Sea Bass/ Barramundi (*Lates calcarifer*): proceedings of an international workshop held at Darwin, N.T., Australia, 24–3- September 1986. Australian Centre for International Agricultural Research (ACIAR), ACIAR Proceedings No. 20, Canberra.

Guindon, K. G. 2011. Evaluating lethal and sublethal effects of catch-and-release angling in Florida's central Gulf Coast recreational Atlantic Tarpon (*Megalops atlanticus*) fishery. Doctoral dissertation. University of South Florida, Tampa.

Haley, V. 2009. Acoustic telemetry studies of Bonefish (*Albula vulpes*) movement around Andros Island, Bahamas: implications for species management. Master's thesis. Florida International University, Miami.

Hanley, N., J. Shogren, and B. White. 2007. Environmental economics in theory and practice. Palgrave, London.

Harrington, R. W., Jr. 1958. Morphometry and ecology of small Tarpon, *Megalops atlantica*, Valenciennes from transitional stage through onset of scale formation. Copeia 1:1–10.

Henry, G. W., and J. M. Lyle. 2003. The national recreational and indigenous fishing survey. Australian Government Department of Agriculture, Fisheries and Forestry, FRDC Project No. 99/158, Canberra.

Jud, Z. R., C. A. Layman, and J. M. Shenker. 2011. Diet of age-0 Tarpon (*Megalops atlanticus*) in anthropogenically-modified and natural nursery habitats along the Indian River Lagoon, Florida. Environmental Biology of Fishes 90:223–233.

Lambeck, R. J. 1997. Focal species: a multispecies umbrella for nature conservation. Conservation Biology 11:849–856.

Layman, C. A., J. P. Quattrochi, C. M. Peyer, and J. E. Allgeier. 2007. Niche width collapse in a resilient top predator following ecosystem fragmentation. Ecology Letters 10:937–944.

Leis, J. M., and S. E. Reader. 1991. Distributional ecology of Milkfish, *Chanos chanos*, larvae in the Great Barrier Reef and Coral Sea near Lizard Island, Australia. Environmental Biology of Fishes 30:395–405.

Ley, J. A. 2005. Linking fish assemblages and attributes of mangrove estuaries in tropical Australia: criteria for regional marine reserves. Marine Ecology Progress Series 305:41–57.

Ley, J. A. 2008. Indo-Pacific Tarpon Megalops cyprinoides: a review and ecological assessment. Pages 3–36 *in* J. S. Ault editor. Biology and management of the world Tarpon and Bonefish fisheries. CRC Press, Boca Raton, Florida.

Ley, J. A., and M. S. Allen. 2013. Modeling marine protected area value in a catch-and-release dominated estuarine fishery. Fisheries Research 144:60–73.

Lotze, H. K., H. S. Lenihan, B. J. Bourque, R. H. Bradbury, R. G. Cooke, M. C. Kay, S. M. Kidwell, M. X. Kirby, C. H. Peterson, and J. B. C. Jackson. 2006. Depletion, degradation, and recovery potential of estuaries and coastal seas. Science 312:1806–1809.

de Menezes, M. F., and M. P. Paiva. 1966. Notes on the biology of Tarpon, *Tarpon atlanticus* (Cuvier and Valenciennes), from coastal waters of Ceara State, Brazil. Arquivos da Estação de Biologia Marinha da Universidade Federal do Ceará 6:83–98.

Mojica R., J. M. Shenker, C. W. Harnden, and D. E. Wanger. 1995. Recruitment of Bonefish, *Albula vulpes*, around Lee Stocking Island, Bahamas. Fishery Bulletin 93:666–674.

Muller, R. G., and R. G. Taylor. 2006. The 2005 stock assessment update of Common Snook, *Centropomus undecimalis*. Florida Marine Fisheries Commission, Florida Marine Research Institute, IHR 2006-003, St. Petersburg.

Mumby, P. J., A. J. Edwards, J. E. Arias-Gonzalez, K. C Lindeman, P. G. Blackwell, A. Gall, M. I. Gorczynska, A. R. Harborne, C. L. Pescod, H. Renken, C. C. C. Wabnitz, and G. Llewellyn. 2004. Mangroves enhance the biomass of coral reef fish communities in the Caribbean. Nature (London) 427:533–536.

Murchie, K. J., S. J. Cooke, A. J. Danylchuk, S. E. Danylchuck, T. L. Goldberg, C. D. Suski, and D. P. Philipp. 2013. Movement patterns of Bonefish (*Albula vulpes*) in tidal creeks and coastal waters of Eleuthera, the Bahamas. Fisheries Research 147:404–412.

Oh, C., and R. B. Ditton. 2006. Using recreation specialization to understand multi-attribute management preferences. Leisure Sciences 28:369–384.

Peters, K. M., R. E. Matheson, Jr., and R. G. Taylor. 1998. Reproduction and early life history of Common Snook, *Centropomus undecimalis* (Bloch), in Florida. Bulletin of Marine Science 62:509–529.

Pelletier, C., K. C. Hanson, and S. J. Cooke. 2007. Do catch-and-release guidelines from state and provincial fisheries agencies in North America conform to scientifically-based best practices? Environmental Management 39:760–773.

Polidoro, B. A., K. E., Carpenter, L. Collins, N. C. Duke, A. M. Ellison, J. C. Ellison, E. J. Farnsworth, E. S. Fernando, K. Kathiresan, N. E. Koedam, S. R. Livingstone, T. Miyagi, G. E. Moore, V. N. Nam, J. E. Ong, J. H. Primavera, S. G. Salmo, III, J. C. Sanciangco, S. Sukardjo, Y. Wang, and J. W. H. Yong. 2010. The loss of species: mangrove extinction risk and geographic areas of global concern. PLOS ONE 5:1–10.

Russell, D. J., and R. N. Garrett. 1988. Movements of juvenile Barramundi, *Lates calcarifer* (Bloch), in north-eastern Queensland. Australian Journal of Marine and Freshwater Research 39:117–123.

Rypel, A., and C. A. Layman. 2008. Degree of aquatic ecosystem fragmentation predicts population characteristics of Gray Snapper (*Lutjanus griseus*) in Caribbean tidal creeks. Canadian Journal of Fisheries and Aquatic Sciences 65:335–339.

Sawynok, W., and J. Pepperell. 2002. National strategy for the survival of released line caught fish planning, project management and communications. Available: http://frdc.com.au/research/final-reports/Pages/2002-099-DLD.aspx (November 2014).

Schlaifer, A., and C. M. Breder. 1940. Social and respiratory behavior of small Tarpon. Zoology 25:493–512.

Seymour, R. S., K. Christian, M. B. Bennett, J. Baldwin, R. M. G. Wells, and R. V. Baudinette. 2004. Partitioning of respiration between the gills and air-breathing organ in response to aquatic hypoxia and exercise in the Pacific Tarpon, *Megalops cyprinoides*. Physiological and Biochemical Zoology 77:760–767.

Sklar, F .H., and J. A. Browder. 1998. Coastal environmental impacts brought about by alterations in freshwater flow in the Gulf of Mexico. Environmental Management 22:547–562.

Staunton-Smith, J., J. B. Robins, D. G. Mayer, M. J. Sellin, and I. A. Halliday. 2004. Does the quantity and timing of fresh water flowing into a dry tropical estuary affect year-class strength of Barramundi (*Lates calcarifer*)? Marine and Freshwater Research 55:787–797.

Stevens, P. W., D. A. Blewett. G. R. Poulakis. 2007. Variable habitat use by juvenile Common Snook, *Centropomus undecimalis* (Pisces: Centropomidae): applying a life-history model in a southwest Florida estuary. Bulletin of Marine Science 80:83–108.

Taylor, R. G., H. J. Grier, and J. A. Whittington. 1998. Spawning rhythms of Common Snook in Florida. Journal of Fish Biology 53:502–520.

Taylor, R. G., J. A. Whittington, H. J. Grier, and R. E. Crabtree. 2000. Age, growth, maturation, and protandric sex reversal in Common Snook, *Centropomus undecimalis*, from the east and west coasts of South Florida. Fishery Bulletin 98:612–624.

Taylor, R. G., J. A. Whittington, and D. E. Haymans. 2001. Catch-and-release mortality rates of Common Snook in Florida. North American Journal of Fisheries Management 21:70–75.

Thirunavukkarasu, A. R., A. Mathew, and K. Kailasam, 2004. Handbook of seed production and culture of Asian Seabass, *Lates calcarifer* (Bloch). CIBA Bulletin 18:1–58.

Tsukamoto, Y., and M. Okiyama. 1997. Metamorphosis of the Pacific Tarpon, *Megalops cyprinoides* (Elopiformes, Megalopidae) with remarks on development patterns in the Elopomorpha. Bulletin of Marine Science 60:23–36.

Tucker, J. W., D. J. Russell, and M. A. Rimmer, 2002. Barramundi culture: a success story for aquaculture in Asia and Australia. World Aquaculture 33:53–59.

Tzeng, W., C. Wu, and Y. Wang. 1998. Age of Pacific Tarpon, *Megalops atlanticus*, at estuarine arrival and growth during metamorphosis. Zoological studies 37:177–183.

Twomey, E., and P. Byrne. 1985. A new record for the Tarpon, *Megalops atlanticus* Valenciennes (Osteichthyes-Elopiformes-Elopidae), in the eastern North Atlantic. Journal of Fish Biology 26:359–362.

Valiela, I., J. L. Bowen, and J. K. York. 2001. Mangrove forests: one of the worlds threatened major tropical environments. Bioscience 51:807–815.

Wade, R. A. 1962. The biology of the Tarpon, *Megalops atlanticus*, and the Ox-eye, *Megalops cyprinoides*, with emphasis on larval development. Bulletin of Marine Science 12:545–599.

Wells, R. M. G., R. S. Baldwin, R. V. Baudinette, K. Christian, and M. B. Bennett. 2003. Oxygen transport capacity in the air-breathing fish, *Megalops cyprinoides*: compensations for strenuous exercise. Comparative Biochemistry and Physiology A 134:45–53.

Whitehead, P. J. P., and R. Vergara. 1978. Megalopidae. In W. Fischer, editor. FAO species identification sheets for fishery purposes. Western Central Atlantic (Fishing Area 31). Volume 3. Food and Agriculture Organization of the United Nations, Rome.

Zerbi, A., C. Aliaume, and J. C. Joyeux. 2001. Growth of juvenile Tarpon in Puerto Rican estuaries. ICES Journal of Marine Science 58:87–95.

American Fisheries Society Symposium 83:57–60, 2015

EXTENDED ABSTRACT

A Preliminary Assessment of Caribbean Reef Fish Abundance in Relation to Mangrove Forest Area

JOSEPH E. SERAFY*
National Marine Fisheries Service, Southeast Fisheries Science Center
75 Virginia Beach Drive, Miami, Florida 33149, USA
and
Rosenstiel School of Marine and Atmospheric Science, University of Miami
4600 Rickenbacker Causeway, Miami, Florida 33149, USA

GEOFFREY S. SHIDELER AND RAFAEL J. ARAÚJO
Rosenstiel School of Marine and Atmospheric Science, University of Miami
4600 Rickenbacker Causeway, Miami, Florida 33149, USA

Scientists have long sought to understand the contribution of mangrove forests to offshore systems, especially the movement of fishes from vegetated shallows to coral reefs (Parrish 1989). Typically, it is on coral reefs that fishing pressure is most intense (Stallings 2009). Previous studies focusing on mangrove-utilizing Caribbean fishes that migrate offshore with ontogeny have been restricted in spatiotemporal extent (e.g., Nagelkerken et al. 2002), and methodological and other differences among these research efforts have hindered quantitative, interstudy comparisons. Moreover, these studies did not directly account for (i.e., incorporate in their data analyses) the potentially confounding influence of human activity (i.e., fishing and/ or habitat degradation) when examining for relationships linking mangrove presence or area to fish abundance on reefs. As a result, it remains unclear whether island-scale observations consistent with mangrove-mediated augmentation of fishes on coral reefs are also evident at the scale of the greater Caribbean region. Human influence on large, high-trophic level Caribbean fishes appears strong (Stallings 2009); therefore, it is important to account for this influence when examining for a mangrove-driven fish subsidy effect at the regional scale.

Here, we draw on an extensive fishery-independent fish abundance data collection effort that spans coral reefs throughout the Caribbean region from Florida, USA south to Venezuela. The REEF (Reef Environmental Education Foundation) Fish Survey is a citizen scientist-based effort whereby REEF personnel perform underwater fish identification training as well as collect, manage, and disseminate reef fish abundance information upon request. Since the effort's 1993 inception, approximately 49,500 fish surveys, performed by experienced volunteer divers using the same roving diver technique, have been logged throughout the Caribbean region. In the present study, we examined variation in the abundances on reefs of 11 fishes (see Table 1 for list of fishes) that have previously been identified as highly or possibly mangrove dependent (Nagelkerken et al. 2002). Specifically, relative abundance of each species was regressed against es-

* Corresponding author: jserafy@rsmas.miami.edu

Table 1.—Regression relationships linking fish abundance with mangrove extent, human population size, and/or latitude. Shown are species-specific final models emerging after (stepwise elimination) multiple regression analyses. MD = mangrove dependence as posited by Nagelkerken et al. (2002), where 1 = highly dependent and 2 = possibly dependent. Fish abundance (A) is predicted as a function of mangrove forest area (M), human population (P), and/or latitude (L). Arrows next to model factors indicate direction of effect (↑ = positive coefficients, ↓= negative). NS = no model terms as statistically significant (i.e., all factor P-values > 0.05). Adj. R^2 = adjusted R^2 values.

MD	Species	Final model	Adj. R^2
1	Bluestriped Grunt *Haemulon sciurus*	$A = M\uparrow$	0.58
1	Schoolmaster *Lutjanus apodus*	$A = M\uparrow + P\downarrow + L\downarrow$	0.61
1	Yellowtail Snapper *Ocyurus chrysurus*	$A = M\uparrow + P\downarrow$	0.49
1	Striped Parrotfish *Scarus iseri*	NS	
2	Yellowfin Mojarra *Gerres cinereus*	NS	
2	White Grunt *Haemulon plumierii*	$A = M\uparrow$	0.61
2	Mutton Snapper *Lutjanus analis*	$A = M\uparrow$	0.32
2	Gray Snapper *Lutjanus griseus*	$A = M\uparrow$	0.30
2	Blue Parrotfish *Scarus coeruleus*	NS	
2	Rainbow Parrotfish *Scarus guacamaia*	$A = M\uparrow$	0.32
2	Great Barracuda *Sphyraena barracuda*	$A = M\uparrow + P\downarrow$	0.47

timates of mangrove forest area (obtained from Spalding et al. 1997), coastal human population (The World Bank, http://data.worldbank.org/indicator/SP.POP.TOTL), and latitude using a stepwise elimination approach. Due to problems of nonnormality and heteroscedasticity, all data (i.e., independent and dependent variables) were rank-transformed prior to applying general linear models (Iman and Conover 1979). The initial regression model took the form

$$A = M + P + L + M{*}P + M{*}L + P{*}L + M{*}P{*}L,$$

where A was mean species-specific fish abundance for each nation, M was mangrove forest area, P was absolute (coastal) human population size, and L was latitude. Beginning with the highest order terms, factors were successively removed if they were not statistically significant ($P > 0.05$) and final model fit was judged from its adjusted R^2 value. Mangrove forest area estimates were obtained from Spalding et al. (2010), and human coastal population estimates from various agency sources on the Internet. Latitudes were those reported in the REEF Fish Survey database (REEF 2014).

Final models emerging from the stepwise regression analyses (Figure 1; Table 1) indicated that (1) interaction effects among independent variables were not significant; (2) for 8 of the 11 species examined, a significant positive mangrove main effect was detected; (3) a human population effect was evident for three species such that all of their abundances were inversely related to population size; (4) latitude emerged significant for just one species (Schoolmaster), whereby fish abundance declined as it increased; and (5) for three species, none of the factors were significant (including mangrove forest area). For those species for which final models emerged, coefficients of determination (adjusted R^2 values) ranged from 0.30 to 0.61.

At the Caribbean scale, our results are not consistent with mangrove dependence for Striped Parrotfish, Blue Parrotfish, and Yellowfin Mojarra, which were listed as either highly or possibly mangrove dependent by Nagelkerken et al. (2002) based on their

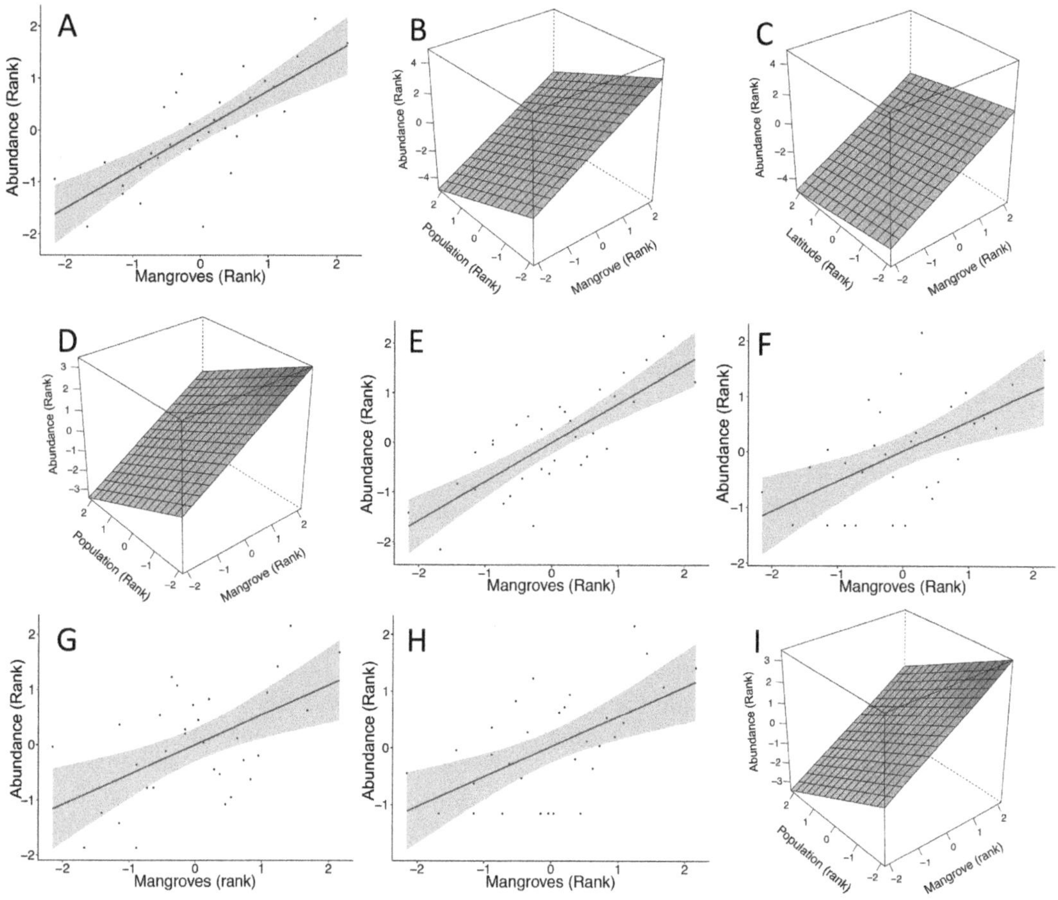

Figure 1.—Graphic depiction of relationships among fish abundance, mangrove extent, human population size, and latitude. Shown are species-specific final models emerging after (stepwise elimination) multiple regression analyses for (A) Bluestriped Grunt, (B) Schoolmaster showing population effect, (C) Schoolmaster showing latitude effect, (D) Yellowtail Snapper, (E) White Grunt (F) Mutton Snapper, (G) Gray Snapper, (H) Rainbow Parrotfish, and (I) Great Barracuda. Panels ordered following the Nagelkerken et al. (2002) posited level of mangrove dependence for each fish species (see Table 1).

island-scale observations. The lack of relationship may reflect that the abundances of these species relate to the quantity of a specific mangrove forest type within the total that cannot be readily discerned at the regional scale. However, our results are consistent with mangrove-to-reef enhancement for the majority (73%) of species on the Nagelkerken et al. (2002) list. For three of these eight, a significant anthropogenic signal (likely reflecting fishing pressure) was superimposed on the positive mangrove effect. Latitude explained little variation in fish abundances, a finding also reported by Stallings (2009) in his analysis of the occurrence of large predatory reef fish in the greater Caribbean region.

In a recent review of linkages among mangrove and sea grass habitats to fisheries, Saenger et al. (2013) warned against generalizing about the importance of mangroves within and among regions. For the greater Caribbean region, our results tend to support the generalization that abundances of several fish species on coral reefs are limited by the size of adjacent mangrove forests. Key to our analysis was taking a species-specific approach and accounting for other factors, especially fishing pressure, which in some cases, would have obscured the underlying, positive influence of mangrove habitat extent. Additional research is needed to account for other confounding effects such as sea grass and reef surface areas. Our study is another of several that underscore the merit of volunteer, citizen-scientist organizations to provide large, valuable data sets that extend well beyond the spatiotemporal scope of most conventional governmental or academic research entities. Clearly, the REEF fish data are valuable for addressing many challenging questions surrounding Caribbean marine fish resources in light of regional human population growth, climate change, and sea level rise.

References

Iman, R. L., and W. J. Conover. 1979. The use of the rank transform in regression. Technometrics 21:499–509.

Nagelkerken, I., C. M. Roberts, G. van der Velde, M. Dorenbosch, M. C van Riel, E. Cocheret de la Morinière, and P. H. Nienhuis. 2002. How important are mangroves and seagrass beds for coral-reef fish? The nursery hypothesis tested on an island scale. Marine Ecology Progress Series 244:299–305.

Parrish, J. D. 1989. Fish communities of interacting shallow water habitats in tropical oceanic regions. Marine Ecology Progress Series 58:143–160.

REEF (Reef Environmental Education Foundation). 2014. Reef Environmental Education Foundation Volunteer Survey Project database. Available: www.reef.org. (February 2014).

Saenger, P., D. Gartside, and S. Funge-Smith. 2013. A review of mangrove and seagrass ecosystems and their linkage to fisheries and fisheries management. Food and Agriculture Organization of the United Nations, Regional Office for Asia and the Pacific, RAP Publication 2013/09, Bangkok, Thailand.

Spalding, M., F. Blasco, and C. Field, editors. 1997. World mangrove atlas. The International Society for Mangrove Ecosystems, Okinawa, Japan. Available: https://ia600507.us.archive.org/10/items/worldmangroveatl97spal/worldmangroveatl97spal.pdf. (January 2015).

Spalding, M., M. Kainuma, and L. Collins. 2010. World atlas of mangroves. Earthscan, London.

Stallings, C. D. 2009. Fishery-independent data reveal negative effect of human population density on Caribbean predatory fish communities. PLOS ONE 4(5):e5333.

American Fisheries Society Symposium 83:61–63, 2015

EXTENDED ABSTRACT

Review of the Decline of Artisanal Fisheries along the Arabian Sea Coast, Pakistan

Muhammad Naeem Khan*
Department of Zoology, University of the Punjab
Quaid-i-Azam Campus (New Campus), Lahore 54590, Pakistan

Pakistan's coastline forming the northern boundary of the Arabian Sea is about 990 km long having 22,820 km2 of territorial waters, which include the 320-km-long Sindh coast bordering with India in the east and the 670-km-long Baluchistan's Makran coast bordering with Iran in the west (Figure 1). While the deep-sea fisheries in Pakistan's exclusive economic zone are managed by the federal government of Pakistan, the 12-nautical-miles coastal zone, reserved for the exclusive use of small-scale artisanal fishers, is the domain of the provincial governments of Sindh and Baluchistan. The paper describes the poor state of artisanal fisheries in Pakistan by giving a short overview of the factors causing the decline with a focus on the rapid deforestation of mangroves in the Indus River delta, belatedly declared a protected Ramsar site.

Most of the Pakistan coastline, except the Sindh coast of the Indus River delta, is arid with scanty clusters of fishing villages and landing sites at Thatha, Badin, Ormara, Pasni, Gwader and Jiwani sea ports. These poor fishing communities are engaged with small-scale artisanal fisheries as natural fish stocks in these coastal waters have drastically declined during the last two to three decades due to mangrove deforestation, overfishing, aquatic pollution, and other anthropogenic activities. As a consequence, almost 79% of these fishing communities now live below the poverty line (Mangroves for the Future 2010). The traditional fishing methods were generally considered environmentally friendly as they did not harm the ecosystem. However, the introduction of new mechanized boats and technologies equipped with nets made of nylon having finer mesh size are becoming harmful to the environment and sustainable fish stocks. The poverty-resource degradation nexus is further contributing to this decline in natural resources, thereby reinforcing the poverty of the artisanal fisher (Khan and Khan 2011).

While the marine fisheries sector in Pakistan directly supports around 125,000 people for both food and income subsistence, almost 1 million people are indirectly dependent upon these marine fisheries resources. These artisanal fishermen usually use small boats between 18 and 25 ft in length to fish in the coastal waters in creeks and within the 12-mi territorial limits (Khan and Khan 2011). During the past 20 years, while the fishing fleet had grown by 15%, the fisheries resources have declined drastically and the fish catches have dropped significantly low. A recent assessment survey along Pakistan's coast carried out over 19 years show a one-third decline in stocks and catches of key fish species like Mangrove Red Snapper *Lutjanus argentimaculatus*, Randall's Threadfin Bream *Nemipterus randalli*, Cobia *Rachycentron canadum*, and Coral Trout *Plectropomus*

* Corresponding author: naeem.zool@pu.edu.pk

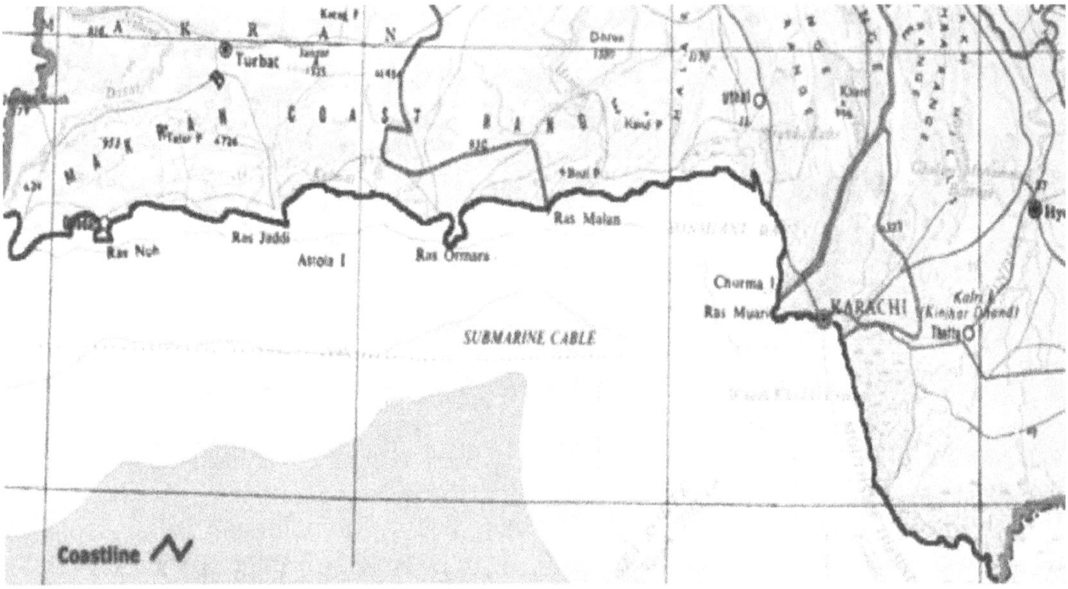

Figure 1.—Map showing the coastline of Pakistan in the Arabian Sea, Indian Ocean.

leopardus (Wijeratna 2007; Khan and Khan 2011).

The mangrove forest along the Indus delta is the largest arid climate mangrove forest of the world. However, they are fast disappearing because of reduction in river flow by dams and barrages (Irum and Hannan 2012). Over the past 50 years, there has been considerable loss of mangrove forest in Pakistan. Mangrove area has reportedly falling from 600,000 ha in 1932 to 1,00,000 ha in 2010 (Memon 2012). It is universally agreed that any disturbance to the mangrove ecosystem will result in smaller fish population and reduction in catches in the off-site fisheries. Therefore, reduction in the mangroves is likely to be one of the major causes of the reduction in artisanal fish stocks as these mangroves provide food and shelter to fish (IUCN 2005). The mangroves act as nurseries and feeding grounds for economically important fish species on which artisanal fisheries communities depends. Of the more than 100 species of fish that have been recorded from the mangroves in Pakistan, almost half of them were in fingerling or young stages while the other half were in subadult or adult stages (IUCN 2005). Recent studies have suggested that productivity of these mangrove ecosystems have been reduced due to the decrease in the nutrients flowing down the River Indus, ultimately resulting in fisheries decline (Irum and Hannan 2012).

Aquatic pollution is also an important reason for the decline of artisanal fisheries in Pakistan. The coastal habitats and the biodiversity are subject to increasing pressures arising from these anthropogenic activities. There are many sources of pollution in the coastal areas, including oil spills, household sewage, industrial waste, and agricultural runoff. Almost 70% of Karachi's wastewater, which is around 350 million gallons, flows into the marine environment untreated. The sewage and agriculture runoff have also resulted in eutrophication in coastal waters leading to harmful algal blooms and fish kills (Khan and Khan 2011).

Institutional and policy shortcomings are also strong reasons for the decline in artisanal fisheries. Fisheries management measures appear to be confined to a few technical management measures, such as closed area and closed seasons. There is no comprehensive policy plan for sustainable fisheries development, management, conservation, and restocking of fingerlings. There is no limit on the number of boats fishing in territorial waters as the number has increased by 15% during the past two decades. Enforcement of fisheries regulations is a neglected aspect of fisheries management in Pakistan (Wijeratna 2007; Khan and Khan 2011). Due to a shortage of funding and trained human resources, fisheries departments and other enforcement agencies are ill-equipped, ineffective, and unable to implement the fishing regulations, thereby promoting overharvesting and the decline of artisanal fisheries in Pakistan.

The lack of environmental awareness, absence of stock assessments, and nonreplenishment of the depleting fish stocks through artificial marine fish hatcheries are further complicating and aggravating the artisanal fisheries in Pakistan. It is encouraging that civil society and community organizations like the International Union for Conservation of Nature, World Wildlife Fund, Fisher Folk, and educational institutions have successfully raised awareness to protect mangroves of the Indus delta, now declared a Ramsar site. Similarly, the community has been involved in mangrove replantation campaigns and against the release of untreated industrial effluents and municipal wastes in marine environment. Therefore, the very survival of artisanal fisheries lies in the change of thought process of the communities, agencies, and other stakeholders, along with the strong political will of the government.

References

Irum, M., and A. Hannan. 2012. Constrains on mangrove forests and conservation projects in Pakistan. Journal of Coastal Conservation 16:51–62.

IUCN (International Union for Conservation of Nature) Pakistan. 2005. Mangroves of Pakistan: status and management. IUCN Pakistan, Karachi, Pakistan.

Khan, S. R., and Khan, S. R. 2011. Fishery degradation in Pakistan: a poverty–environment nexus? Canadian Journal of Development Studies 32:32–47.

Memon, S. H. 2012. An overview of mangrove restoration efforts in Pakistan. Pages 51–61 *in* Sharing lessons on mangrove restoration: proceedings and a call for action from an MFF regional colloquium 30–31 August 2012, Mamallapuram, India. IUCN, Gland, Switzerland with Mangroves for the Future, Bangkok, Thailand.

Mangroves for the Future. 2010. Pakistan: national strategy and action plan for Mangroves for the Future. Mangroves for the Future. Available: www.mangrovesforthefuture.org/assets/Repository/Documents/MFF-Pakistan-NSAP2.pdf. (March 2015).

Wijeratna, A. 2007. Taking the fish: fishing communities lose out to big trawlers in Pakistan. ActionAid International, Johannesburg, South Africa.

Community Ecology and Connectivity

American Fisheries Society Symposium 83:67–80, 2015

Fish Community and Habitat Assessments of Three Adjacent Tidal Creeks on Cape Eleuthera, The Bahamas

Karen J. Murchie*
Department of Biology, College of The Bahamas, Grand Bahama Highway Post Office Box F-42766, Freeport, Grand Bahama, The Bahamas

Sascha Clark Danylchuk
Fish Mission
11 Kingman Road, Amherst, Massachusetts 01002, USA

Andy J. Danylchuk
Department of Environmental Conservation, University of Massachusetts 311 Holdsworth Hall, 160 Holdsworth Way, Amherst, Massachusetts 01003, USA

Steven J. Cooke
Fish Ecology and Conservation Physiology Laboratory
Department of Biology and Institute of Environmental Science, Carleton University 1125 Colonel By Drive, Ottawa, Ontario K1S 5B6, Canada

Abstract.—Three adjacent tidal creek systems (Page, Kemps, and Broad creeks) on Cape Eleuthera, The Bahamas were studied to quantify the variation in fish community structure and habitat characteristics over small (<5 km) spatial scales. Snorkeling transects were used to census the fish community on a summer new moon during slack high tide and involved the simultaneous assessment of each creek and each zone within the creek (i.e., mouth, middle, and upper) replicated over three consecutive days. The simultaneous assessment (involving large teams) was done to enable direct comparison without spatial sampling being confounded by time. Habitat assessments included measurements of water quality parameters, sediment sampling, and vegetation surveys. Despite their close proximity, creeks differed in both fish community structure and habitat characteristics. Broad Creek had the greatest fish species richness (n = 15), followed by Kemps Creek (n = 14) and Page Creek (n = 10). Mangrove habitats had significantly greater fish species diversity in Broad Creek while sea grass habitats resulted in higher species richness in Page Creek, relative to other habitat types. Mangrove and algal plain habitats had the highest fish species diversity in Kemps Creek. Within creeks, fish abundance was dependent on zonation, with the largest number of fish being found in creek mouths compared to upper sections. Water quality parameters (i.e., temperature, dissolved oxygen, and salinity) differed among the creeks, presumably reflecting creek morphology. Out of the 10 different species of vegetation observed, 60% were found in all tidal creeks. Coarse sand was the predominant particle size for all creeks, with

* Corresponding author: karen.murchie@gmail.com

variation in the second most abundant particle size between Page Creek and the others. This study reveals the great heterogeneity of tidal creek fish community and habitat characteristics and illustrates that conservation and management strategies along with monitoring programs must recognize the variation that can occur among and within coastal creeks over relatively small spatial scales.

Introduction

Coastal creek systems are an important nearshore seascape, comprised of functionally connected habitats, including sea grasses, algal plains, sand flats, soft-sediment beds, and mangroves (Krumme 2009; Boström et al. 2011). Traditionally, habitats comprising this seascape have been studied as independent entities within the context of a larger coastal ecosystem. For example, mangroves (Blaber 2007) and sea grass meadows (Heck et al. 2003) have received much attention from scientists on an individual basis because of their high productivity, ability to stabilize sediment, and provision of key habitats for fish and invertebrate taxa. Mangroves in particular function as nursery, foraging, and refuge areas for fish in subtropical and tropical regions (Mumby et al. 2004; Lugendo et al. 2006; Nagelkerken et al. 2008; Walters et al. 2008). Due to the regular submersion and exposure of mangrove habitats associated with tidal cycles in most subtropical and tropical areas, however, few fish can exclusively inhabit these patches. Rather, some fish must migrate in and out, using alternative habitats such as sea grass beds when mangroves are unavailable (Sheaves 2005). Thus, the habitats comprising the coastal creek system mosaic are inherently connected not only through the physiochemical processes associated with tides, but also through the biota that move between the patches (Moberg and Folke 1999; Mumby 2006; Hitt et al. 2011).

At present, one of the most pressing needs of resource managers and decision makers is a better understanding of ecological connectivity in subtropical and tropical coastal systems (Grober-Dunsmore et al. 2009). Given the myriad of anthropogenic modifications that have occurred in these areas (reviewed in Gladstone 2009), a global push for conservation and management of subtropical and tropical coastal systems is crucial to reduce, reverse, and prevent unnatural changes and to maintain the ecological goods and services that they provide (UNEP 2006). While creating marine protected areas is one management strategy, it has been highlighted that current assessment practices of nursery habitats, for example, often take a static approach of considering coastal creek systems as individual and homogeneous entities (Nagelkerken et al. 2015). Instead, such seascapes are dynamic and spatially heterogeneous, with some configurations resulting in increased connectivity (Grober-Dunsmore et al. 2009) and higher densities of fish communities (MacDonald and Weis 2013).

To date, research efforts on coastal tidal creek systems tend to be focused on a single system or across multiple systems covering a broad geographic range. There are relatively few studies that examine the variation in fish assemblages and habitat in adjacent creeks on small (<5 km) spatial scales. Additionally, many studies at the fish community level (e.g., Layman et al. 2004; Mwandya et al. 2010) are conducted sequentially at different sites, which makes it difficult to control for temporal variability, even across several days where tides, photoperiod, lunar phase, and weather can vary substantially. As such, the purpose of this study was to simultaneously survey the fish community of three tidal creeks along a less than 5-km stretch of coastline on Cape Eleuthera, The Bahamas, replicated over a 3-d period, as well as collect data on the habitat characteristics in each creek.

Methods

Study location

This study took place on the island of Eleuthera, The Bahamas (N 24°50'05" and W 76°20'32"), in three coastal tidal creeks (Page Creek, Kemps Creek, and Broad Creek; see Murchie et al. 2013 for more details). Assessments of habitat and fish assemblages were facilitated by dividing each creek into three zones (mouth, middle, and upper; Figure 1). The mouth zone encompassed the opening of the creek and had the greatest depth at high tide (~1.2 m) and the highest water flow. The upper zone included areas of the creek with the lowest water depth at high tide (~0.3 m), and had little to no water at

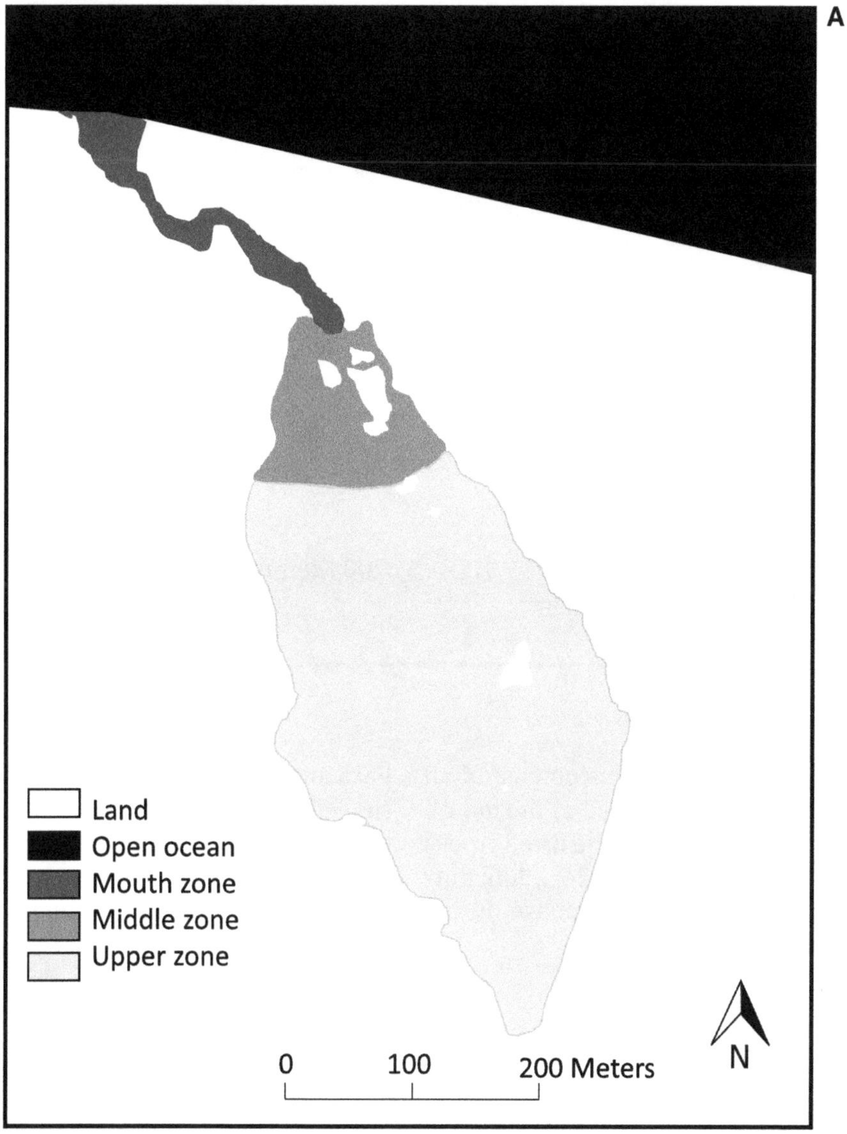

Figure 1.—Habitat zone designations (mouth, middle, and upper) for (A) Page Creek, (B) Kemps Creek, and (C) Broad Creek.

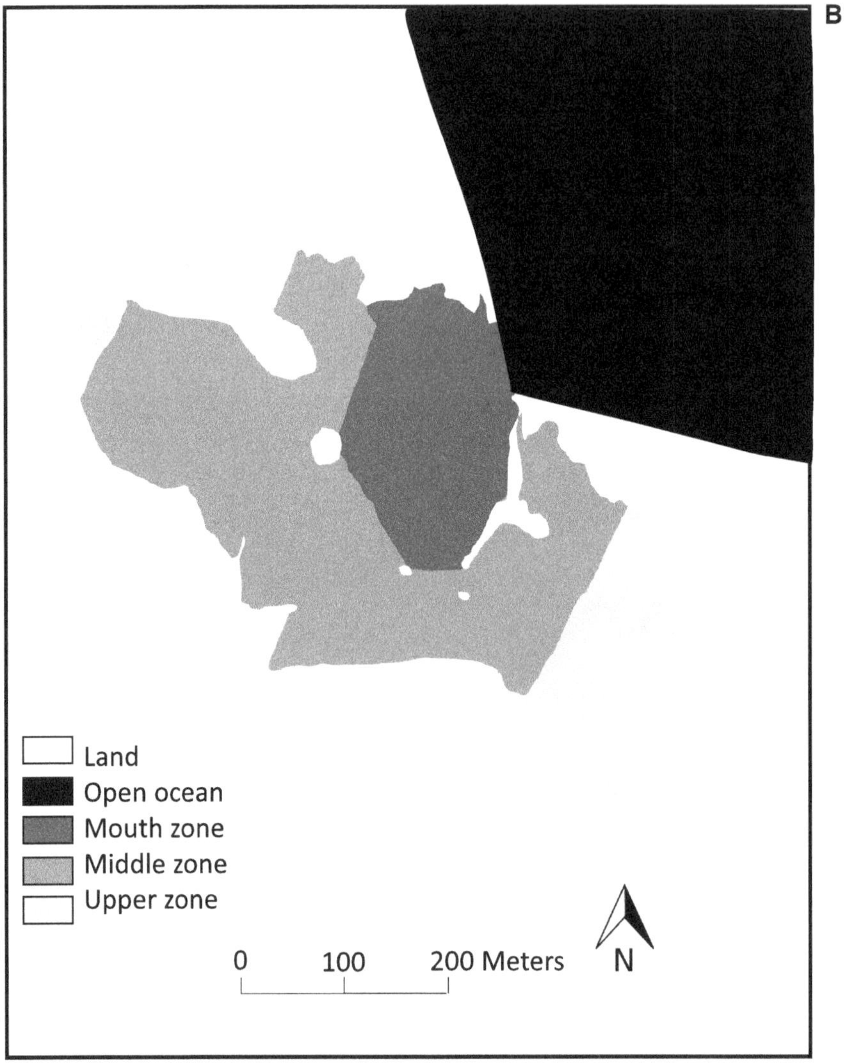

Figure 1.—Continued.

low tide. The middle zone encompassed all areas that were not designated as the mouth or upper zones. Sampling of the three creeks took place from August 22 to 24, 2006, during slack high tide. The lunar phase during this time was a new moon.

Fish community assessment

Underwater visual census was used to assess fish assemblages (Layman et al. 2004). Specifically, pairs of snorkelers swam in each zone of each creek simultaneously to provide a snapshot of the fish communities. Each snorkeling pair swam together in an s-shaped pattern at a rate of 3 m/min for a total of 5 min in each habitat type. All habitats in a specific zone were surveyed three times in a different location by the pair of swimmers. Data from each snorkeling pair was averaged to yield what was considered one sighting. Water depth was less than 2 m for all areas sampled, and visibility was at least 5 m. Species of *Eucinostomus* were lumped together due to the difficulty of accurately identifying Bigeye Mojarra *E. havana*, Flagfin Mojarra *E. melanopterus*, and Mottled Mo-

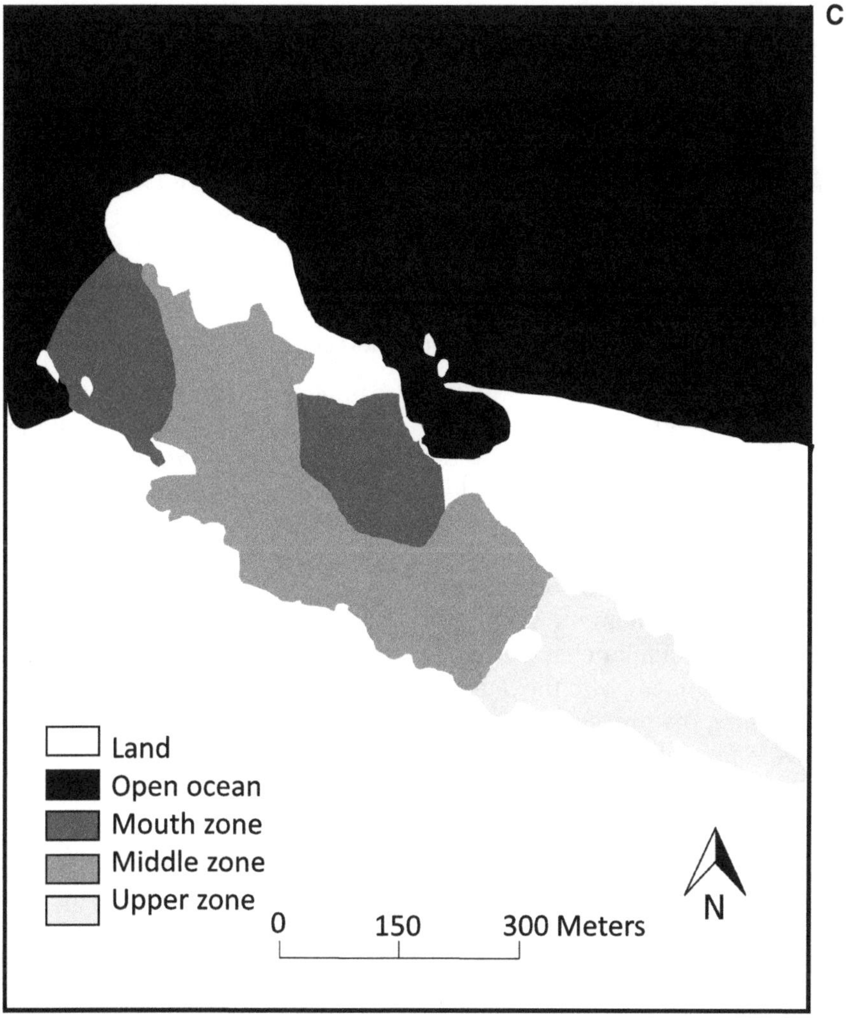

Figure 1.—Continued.

jarra *E. lefroyi* under water. Other fish were identified to the lowest taxonomic level possible (e.g., damselfishes *Stegastes* spp., gobies *Gobidae* spp., needlefishes *Strongylura* spp., and silversides [atherinid species]). Species richness and relative abundance were determined for each creek and were summarized with a rank-abundance curve. A two-way analysis of variance (ANOVA) was used to test if there was a difference in the abundance of fish by zone within creek and among creeks. For each creek, an ANOVA was performed to examine if there was a temporal difference (over the 3 d of sampling) in fish abundance and diversity, as well as to assess if there were differences in fish abundance and diversity by habitat. Tukey-Kramer honest significance difference tests were performed following ANOVAs to determine where significant differences lie between variables (Day and Quinn 1989).

Habitat assessment

Water temperature, dissolved oxygen, and salinity were measured at increasing distances from the mouth of each creek using a YSI water quality meter (YSI 85, YSI Inc., Yellow Springs, Ohio, USA). An average of

12 readings were taken per creek up to a distance of 515 m from the mouth. Regression analyses were preformed to determine whether a relationship existed between the water-quality parameters and distance from the creek mouth. An ANOVA was used to determine whether there were differences in water temperature, salinity, and dissolved oxygen between the three coastal creeks.

Within each creek, habitat types were assessed. Habitat designations were based on Layman et al. (2004). Sand flats were defined by a rippled sandy bottom, as well as a lack of any type of dense vegetation. Algal flats had more than 50% algae cover while the sea grass habitat had greater than 50% sea grass cover. Soft sediment was defined as any area in which compaction of sediment was so low that experimenters sank. The mangrove habitat had similar characteristics to that of the soft sediment with the one major difference being the presence of mature red mangrove *Rhizophora mangle* or black mangrove *Avicennia germinans* trees. Within each zone (i.e., mouth, middle, and upper), vegetation sampling was accomplished by haphazardly throwing a bobbered weight and sampling the area where it landed. A 1 × 1 m quadrat was used to standardize the sampling area. Percentage of vegetation cover, along with a list of each species present, was recorded. A total of 10 quadrats were sampled in each zone of each creek.

A total of three sediment samples were collected per each habitat type found in each of the tidal creeks. Surface sediment samples were collected by filling a 200-mL cup to capacity. Following collection, samples were sieved through four mesh screens of decreasing size (#5, #10, #35, #120 U.S. standard sieve mesh size). The percentage composition by grain size (Wentworth 1922) was determined following sieving. A two-way ANOVA was conducted to determine if there was any difference in the overall percentage of sediment composition by creek and by habitat. As above, Tukey-Kramer honest significant difference tests were performed following all ANOVAs to determine where significant differences lie between variables. All statistical analyses on collected and derived data were completed using JMP 10 (SAS Institute, Cary, North Carolina). Maximal type-1 error rates were set at $\alpha = 0.05$.

Results

Fish community assessment

A total of 10,421 individual fish were documented during the 3-d simultaneous assessment of all three zones in Page, Kemps, and Broad creeks, with the greatest abundance of fish being recorded in Kemps Creek (n = 4,479; Table 1). Broad Creek had the next highest abundance of fish recorded (n = 4,022) followed by Page Creek (n = 1920). Rank-abundance curves for the three creeks showed that Broad Creek had the highest species richness (n = 15) followed by Kemps Creek and Page Creek with 14 and 10 species, respectively (Figure 2). Broad Creek had the most gradual slope of the rank-abundance curves, indicating a more equitable distribution of individuals among the species, which is a measure of species evenness. Out of the 15 species observed, 66% were documented in all three of the coastal creeks, with 47% of the species being documented in all three of the zones (i.e., mouth, middle, and upper; Table 1). All mojarra species (*Eucinostomus* spp. and Yellowfin Mojarra *Gerres cinereus*), as well as needlefish *Strongylura* spp. and silversides (atherinid species) were found in each of the five habitat types (Table 1).

A two-way ANOVA revealed no difference in the abundance of fish by creek (F = 1.93, df = 2, P = 0.15), but there was a significant difference in the abundance of fish by zone (F = 6.17, df = 2, P = 0.002). In particular, the mouth of the creeks contained a significantly larger number of fish, on average, than the upper portions of the creeks. No temporal variation in fish abundance or fish diversity was detected for any of the tidal creeks (all P-

Table 1.—Summary of the total number of each fish species observed

		Number of fish observed in each creek				
Common name	Scientific name	Page Creek (n = 1,920)	Kemps Creek (n = 4,479)	Broad Creek (n = 4,022)	Zone[a]	Habitat[b]
Great Barracuda	*Sphyraena barracuda*	15	24	7	1, 2	A, M, F, G
Bluehead	*Thalassoma bifasciatum*	0	0	2	1	G
Checkered Puffer	*Sphoeroides testudineus*	4	12	7	1, 2, 3	A, M, F, G
Damselfishes	*Stegastes* spp.	3	7	12	1, 2	A, M, G
French Grunt	*Haemulon flavolineatum*	828	441	1,054	1, 2	A, M, G
Gobies	Gobiidae	3	13	1	1, 2, 3	A, M, G
Lemon Shark	*Negaprion brevirostris*	2	6	2	1, 2, 3	A, M, G
Mojarras	*Eucinostomus* spp.	2	1	1	1, 2, 3	A, M, F, G, S
Needlefishes	*Stronglylura* spp.	288	201	193	1, 2, 3	A, M, F, G, S
Schoolmaster	*Lutjanus apodus*	337	95	70	1, 2, 3	A, M, F, G
Sergeant Major	*Abudefduf saxatilis*	0	2	4	1, 2	M
Silversides	atherinid species	0	3,433	1,930	1, 2	A, M, F, G, S
Slippery Dick	*Halichoeres bivittatus*	0	13	0	1	A, M, G
Southern Stingray	*Dasyatis americana*	0	0	3	1, 3	A, M
White Mullet	*Mugil curema*	0	58	682	1, 2, 3	A, F, G
Yellowfin Mojarra	*Gerres cinereus*	438	173	54	1, 2, 3	A, M, F, G, S

[a] Zones in which species were observed: 1 = mouth, 2 = middle, and 3 = upper.

[b] Habitats in which species were observed: A = algal plain, M = mangrove, F = sand flat, G = sea grass, and S = soft sediment.

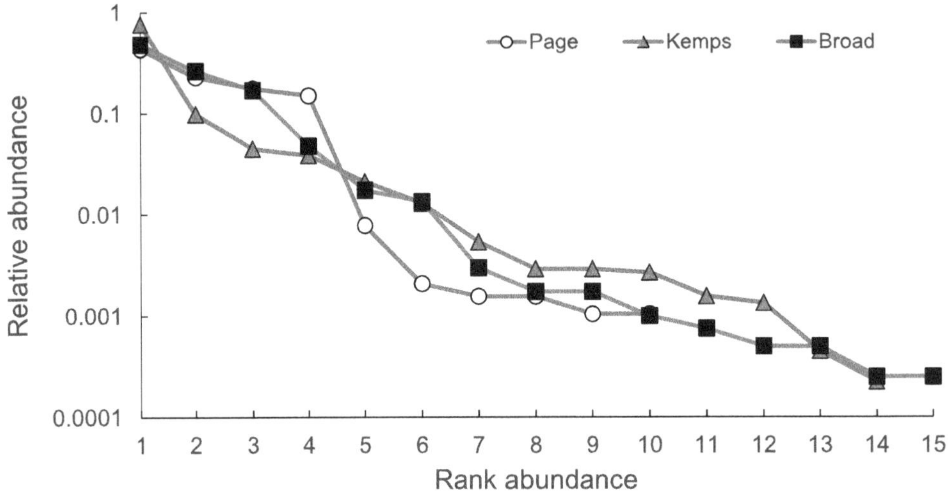

Figure 2.—Rank-abundance curves for the fish communities of Page, Kemps, and Broad creeks.

values > 0.10). There was also no significant difference in the abundance of fish by habitat type for each creek (all *P*-values > 0.20). There was, however, a difference in fish diversity by habitat type for each creek (Page Creek F = 4.61, df = 3, P = 0.005; Kemps Creek F = 8.41, df = 4, P < 0.0001; Broad Creek F = 5.15, df = 3, P = 0.003). Specifically, sea grass habitats had the greatest diversity of fish in Page Creek compared to all other habitat types, whereas mangrove habitats had the greatest diversity of fish in both Broad and Kemps creeks (although algal plain was not significantly different than mangrove habitats in Kemps Creek).

Habitat assessment

Water temperature increased significantly with increasing distance from the mouth zone in Page Creek (r^2 = 0.72, P < 0.001), but there was no significant relationship between water temperature and distance from the mouth for Kemps Creek (r^2 = 0.08, P = 0.49) and Broad Creek (r^2 = 0.04, P = 0.50). Results for salinity varied, with salinity increasing with distance from the mouth zone in Page Creek (r^2 = 0.42, P = 0.001), not changing in Kemps Creek (r^2 = 0.007, P = 0.85), and decreasing with distance from the mouth zone in Broad Creek (r^2 = 0.38, P = 0.015). Dissolved oxygen values increased with increasing distance from the mouth zone in Page and Broad creeks (r^2 = 0.69, P < 0.001 and r^2 = 0.50, P < 0.01, respectively) but did not change significantly in Kemps Creek (r^2 = 0.37, P = 0.107). Water temperature did not vary significantly between the creeks (F = 1.92, df = 2, P = 0.162; Table 3). Salinity was overall lower in Kemps Creek compared to the other two creeks (F = 12.26, df = 2, P < 0.001), and dissolved oxygen was significantly lower in Page Creek, relative to Kemps and Broad Creek (F = 13.94, df = 2, P < 0.0001; Table 2).

During vegetation surveys, a total of 10 species were documented, with five types of algae, three types of sea grass, and two species of mangrove (Table 3). A total of 60% of documented vegetation could be found in all three coastal creeks, with 40% being found in all three zones (*Dasycladus* spp., *Penicillus* spp., *Halodule beaudettei*, and red mangrove; Table 3). Only *Dasycladus* spp. and *H. beaudettei* could be found in all five habitats (Table 3).

For all three coastal creeks, coarse sand was the dominant sediment particle size (Figure 3). Fine sand was the second most abundant sediment particle in Kemps and Broad Creek, but granule was the second most abundant sediment particle in Page Creek. Indeed, Page Creek had a significantly

Table 2.—Mean water quality values (temperature, salinity, and DO) from Page Creek, Kemps Creek, and Broad Creek, in Eleuthera, The Bahamas. Dissimilar letters in each column indicate a significant difference between the creeks following a one-way analysis of variance.

Water quality parameters	Page Creek	Kemps Creek	Broad Creek
Mean temperature (°C)	29.7 z	29.4 z	29.6 z
Mean salinity (parts per thousand)	38.5 z	35.0 y	38.2 z
Mean DO (mg/L)	4.8 y	6.6 z	5.8 z

higher percentage of granule sediment particles than the other two creeks ($F = 5.90$, df = 2, $P = 0.007$), and Kemps and Broad creeks had significantly higher amounts of fine sand compared to Page Creek ($F = 9.71$, df = 2, $P = 0.0006$). The percentage of sediment type by habitat was significantly different for coarse sand ($F = 6.44$, df = 4, $P = 0.0007$) and fine sand ($F = 20.91$, df = 4, $P < 0.0001$). Fine sand was most abundant in mangrove habitats, whereas coarse sand was most comparably abundant in sand flat, sea grass, and algal plains relative to other habitat types.

Discussion

To our knowledge, this is the first study that has conducted simultaneous assessments of fish assemblages in multiple creek systems. While this methodology requires a large number of individuals who are trained to identify the various fish species, it allows the investigators to control for temporal variability where tides, photoperiod, and weather can vary substantially. Additionally, by conducting surveys in each zone of the creek simultaneously, the possibility of recounting the same individual fish is reduced but not eliminated. Although no differences in fish abundance or diversity were detected over the three consecutive day sampling period, longer temporal sampling would be expected to yield changes due to juvenile recruitment and seasonal migration patterns. Underwater visual census techniques such as those that have been employed in this

Table 3.—Summary of the various species of vegetation found during quadrat surveys of three coastal creeks in Eleuthera, The Bahamas.

Category	Scientific name	Creek[a]	Zone[b]	Habitat[c]
Algae	*Dasycladus* spp.	P, K, B	1, 2, 3	A, M, F, G, S
Algae	*Penicillus* spp.	K, B	1, 2, 3	A, F, G
Algae	*Halimeda* spp.	K, B	1, 2	F, G
Algae	*Udotea* spp.	B	2	A
Algae	*Acetabularia* spp.	P, K, B	1, 2	A, F, G
Sea grass	*Thalassia testudinum*	P, K, B	1, 2	A, F, G
Sea grass	*Halodule beaudettei*	P, K, B	1, 2, 3	A, M, F, G, S
Sea grass	*Syringodium filiforme*	P, B	1, 2	G
Mangrove	*Rhizophora mangle*	P, K, B	1, 2, 3	M, F
Mangrove	*Avicennia germinans*	P, K, B	2, 3	A, M

[a] Creeks in which species were observed: P = Page Creek, K = Kemps Creek, and B = Broad Creek.
[b] Zones in which species were observed: 1 = mouth, 2 = middle, and 3 = upper.
[c] Habitats in which species were observed: A = algal plain, M = mangrove, F = sand flat, G = sea grass, S = soft sediment.

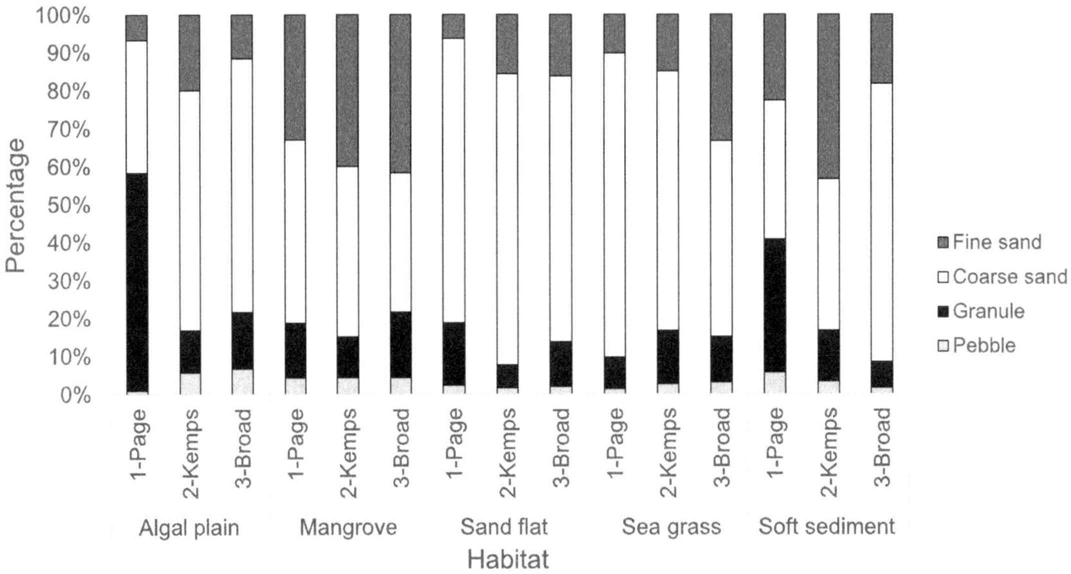

Figure 3.—Percentage of sediment particle size by habitat type for each of the three coastal creeks. The size ranges for the various classes of sediment particle size are as follows: pebble (4–8 mm), granule (2–4 mm), coarse sand (0.5–1 mm), and fine sand (125–250 μm).

study are appealing in the sense that they are nondestructive, but they have been criticized for underestimating fish density and diversity due to fish size (Willis 2001), fish behavior (Kulbicki 1998), and survey procedures (Ward-Paige et al. 2010). Indeed, our survey did not document the presence of Bonefish *Albula vulpes* in any of these tidal creeks, while telemetry studies by Murchie et al. (2013) demonstrate that adult Bonefish use all of these locations daily. It is likely that the activity associated with the large number of snorkelers in the creeks resulted in fish avoiding these areas (Murchie et al. 2009). However, Pita et al. (2014) found that use of human observers is more precise and complete compared to use of remotely operated vehicles or remote underwater video surveys.

It is likely that habitat characteristics and creek morphology are responsible for many of the findings in the fish community survey. For example, fish abundance was highest in creek mouths compared to upper reaches. This can be explained by the fact that habitat characteristics such as depth, water temperature, and salinity tend to make creek mouths more accessible to a wider diversity in size and species of fish (Tzeng and Wang 1992). Also, fish diversity among the creeks was overall highest in mangrove and sea grass habitats relative to soft sediment, sand flats, and algal plains. Mangrove habitats are known to offer many benefits to fish, including feeding grounds, predator refuges, and nursery habitats (Kathiresan and Bingham 2001). The heterogeneous environment of mangroves and their epibiotic diversity has a direct influence on fish community diversity (MacDonald and Weis 2013). Sea grasses are also known to offer many of the same benefits to fish as mangroves and are also considered important nursery habitats (Nagelkerken et al. 2001). One area that deserves attention in the future is the arrangement of the various habitats in these systems (i.e., habitat proximity and the amount of unsuitable low structure patches [e.g., sand flats]) to get a better grasp of ecological connectivity by using a landscape ecology approach (see Grober-Dunsmore et al. 2009; Nagelkerken et al. 2015).

Assessments of water quality parameters, vegetation, and sediment demonstrated the great heterogeneity of these tidal creeks, both within and among systems. While water temperature did not vary overall among the creeks, there were differences in water temperature profiles going from the mouth to the upper portions of each creek. Page Creek was the only location with a significant increase in water temperature with increasing distance from the mouth. This observation may be due to the fact that this creek is smaller in size compared to Kemps and Broad creeks and also because the mouth of Page Creek is narrower, deeper, and mangrove-lined, which blocks some of the direct sunlight. Salinity also increased with increasing distance from the mouth in Page Creek, likely due to evaporation of water from the upper portions of the creek (Sumner and Belaineh 2005). The lack of a salinity gradient from the mouth to upper reaches of Kemps Creek is possibly due to the fact that there are numerous deeper channels that run the length of the creek. The observed decrease in salinity with increasing distance from the mouth of Broad Creek is likely due to its morphology in that there are two mouths with direct exposure to open ocean. Dissolved oxygen increased with increasing distance from the mouths of Page and Broad creeks but did not change significantly in Kemps Creek. Further investigations are required to understand why Page and Broad creeks had increasing levels of dissolved oxygen towards the upper portions of the creek, given that mangrove habitats are typically lower in dissolved oxygen (Lewis and Gilmore 2007).

Vegetation differences that were detected among creeks may have been a result of the random sampling technique employed, and future studies within these systems should census the entire community. *Dasycladus* spp. and *Halodule beaudettei* were the only algae and sea grass, respectively, that were found in all five habitats in the creeks. *Halodule beaudettei* is known to be a pioneering species with an ability to establish productive stands in nutrient-poor sediments and waters (Smith 1996), which can explain its distribution throughout the creeks. The presence of macrophytes, such as sea grasses and macroalgae, is important in providing structure for fish habitat (Mwandya et al. 2010) and can also influence fish predation success (see Jaxion-Harm and Speight 2012).

All three coastal creeks studied were comprised of a heterogeneous mixture of sediment particle sizes, which is common for most coastal systems worldwide (Holland and Elmore 2008). Indeed, heterogeneity is significant in terms of its impact on coastal processes, such as wave friction, sediment transport, and bathymetric change (Holland and Elmore 2008). The bulk of the sediment material was coarse and fine sand, which is made exclusively by carbonate grains (Rankey and Reeder 2012). Mangrove and soft sediment habitats contained the highest concentration of fine sands, which can be attributed to their ease of travel in the water, allowing them to be deposited in middle to upper reaches in the creeks (Collinson and Thompson 1989). Page Creek had granule (2–4 mm) sediment as the second most abundant particle size, whereas both Kemps and Broad creeks had fine sand as the second most abundant particle size. The difference may be explained by the morphology of the creeks. Both Kemps and Broad creeks have much wider mouths (Broad Creek has two openings) when compared to Page Creek. The narrow mouth of Page Creek results in a channel, which concentrates wave energy and allows larger particles such as granules to travel further into the creek as they can only be carried in higher energy water (Collinson and Thompson 1989).

In conclusion, this study revealed the great heterogeneity of tidal creek fish communities and habitat characteristics and illustrated that conservation and management strategies, along with monitoring programs, must recognize the variation that can occur among and within coastal creeks over relatively small spatial scales. In particular, this has implications for site selection of marine protected areas as the heterogeneous nature of each creek and the specific arrangement of habitats will affect not only the diversity of the fish community present, but also its value for each species. While creek morphology clearly plays an important role in contributing to fish abundance and diversity, this is one of the most frequently altered components via anthropogenic disturbance. Managers need to consider all of these factors when confronted with conflicting objectives for conservation and alternative uses of tidal creeks when determining priorities.

Acknowledgments

Alaina Baker, Stephanie Chong, Lisa Duke, Amber Elzner, Lynn Gowman, Alison Kealey, Laura Keeting, Martha Mierzejewski, Viktoria Mileva, Rhea Moggach, Nerissa Podolinsky, Liz Ross, Callie Stanley, and Susan Usjak are acknowledged for their participation in field data collection as part of the Ontario Universities Field Course Program. We also thank the Cape Eleuthera Institute staff for logistical assistance for this project.

References

Blaber, S. J. M. 2007. Mangroves and fishes: issues of diversity, dependence, and dogma. Bulletin of Marine Science 30:457–472.

Boström, C., S. J .Pittman, C. Simenstad, and R. T. Kneib. 2011. Seascape ecology of coastal biogenic habitats: advances, gaps, and challenges. Marine Ecology Progress Series 427:191–217.

Collinson, J. D., and D. B. Thompson. 1989. Sedimentary structures, 2nd edition. Chapman and Hall, London.

Day, R. W., and G. P. Quinn. 1989. Comparisons of treatments after an analysis of variance in ecology. Ecological Monographs 59:433–463.

Gladstone, W. 2009. Conservation and management of tropical coastal ecosystems. Pages 565–605 *in* I. Nagelkerken, editor. Ecological connectivity among tropical coastal ecosystems. Springer, New York.

Grober-Dunsmore, R., S. J. Pittman, C. Caldow, M. S. Kendall, and T. K. Frazer. 2009. A landscape ecology approach for the study of ecological connectivity across tropical marine seascapes. Pages 493–530 *in* I. Nagelkerken, editor. Ecological connectivity among tropical coastal ecosystems. Springer, New York.

Heck, K. L., G. Hays, and R. J. Orth. 2003. Critical evaluation of the nursery role hypothesis for seagrass meadows. Marine Ecology Progress Series 253:123–136.

Hitt, S., S. J. Pittman, and R. S. Nemeth. 2011. Diel movements of fishes linked to benthic seascape structure in a Caribbean coral reef ecosystem. Marine Ecology Progress Series 427:275–291.

Holland, K. T., and P. A. Elmore. 2008. A review of heterogenous sediments in coastal environments. Earth-Science Reviews 89:116–134.

Jaxion-Harm, J., and M. R. Speight. 2012. Algal cover in mangroves affects distribution and predation rates by carnivorous fishes. Journal of Experimental Marine Biology and Ecology 414–415:19–27.

Kathiresan, K., and B. L. Bingham. 2001. Biology of mangroves and mangrove ecosystems. Advances in Marine Biology 40:81–251.

Krumme, U. 2009. Diel and tidal movements by fish and decapods linking tropical coastal ecosystems. Pages 271–324 *in* I. Nagelkerken, editor. Ecological connectivity among tropical coastal ecosystems. Springer, New York.

Kulbicki, M. 1998. How the acquired behaviour of commercial reef fishes may influence the results obtained from visual censuses. Journal of Experimental Marine Biology and Ecology 222:11–30.

Layman, C. A., D. A. Arrington, R. B. Langerhans, and B. R. Silliman. 2004. Degree of fragmentation affects fish assemblage structure in Andros Island (Bahamas) estuaries. Caribbean Journal of Science 40:232–244.

Lewis, R. R., and R. G. Gilmore. 2007. Important considerations to achieve successful mangrove forest restoration with optimum fish habitat. Bulletin of Marine Science 80:823–387.

Lugendo, B. R., I. Nagelkerken, G. van der Velde, and Y. D. Mgaya. 2006. The importance of mangroves, mud and sand flats, and seagrass beds as feeding areas for juvenile fishes in Chwaka Bay, Zanzibar: gut content and stable isotope analyses. Journal of Fish Biology 69:1639–1661.

MacDonald, J. A., and J. S. Weis. 2013. Fish community features correlate with prop root epibionts in Caribbean mangroves. Journal of Experimental Marine Biology and Ecology 441:90–98.

Moberg, F., and C. Folke. 1999. Ecological goods and services of coral reef ecosystems. Ecological Economics 29:215–233.

Mumby, P. J. 2006. Connectivity of reef fish between mangroves and coral reefs: algorithms for the design of marine reserves at seascape scales. Biological Conservation 128:215–222.

Mumby, P. J., A. J. Edwards, J. E. Arias-Gonzalez, K. C. Lindeman, P. G. Blackwell, A. Gall, M. I. Gorczynska, A. R. Harborne, C. L. Pescod, H. Renken, C. C. C. Wabnitz, and G. Llewellyn. 2004. Mangroves enhance the biomass of coral reef fish communities in the Caribbean. Nature (London) 427:533–536.

Murchie, K. J., S. J. Cooke, A. J. Danylchuk, S. E. Danylchuck, T. L. Goldberg, C. D. Suski, and D. P. Philipp. 2013. Movement patterns of Bonefish (*Albula vulpes*) in tidal creeks and coastal waters of Eleuthera, The Bahamas Fisheries Research 147:404–412.

Murchie, K. J., S. E. Danylchuk, C. E. Pullen, E. Brooks, A. D. Shultz, C. D., Suski, A. J. Danylchuk, and S. J. Cooke. 2009. Strategies for the capture and transport of Bonefish, *Albula vulpes*, from tidal creeks to a marine research laboratory for long-term holding. Aquaculture Research 40:1538–1550.

Mwandya, A. W., M. Gullström, M. H. Andersson, M. C. Öhman, Y. D. Mgaya, and I. Bryceson. 2010. Spatial and seasonal variations of fish assemblages in mangrove creek systems in Zanzibar (Tanzania). Estuarine Coastal and Shelf Science 89:277–286.

Nagelkerken, I., S. J. M. Blaber, S. Bouillon, P. Green, M. Haywood, L. G. Kirton, J.-O. Meynecke, J. Pawlik, H. M. Penrose, A. Sasekumar, and P. J. Somerfield. 2008. The habitat function of mangroves for terrestrial and marine fauna: a review. Aquatic Botany 89:155–185.

Nagelkerken, I., S. Kleijnen, T. Klop, R. A. C. J. van den Brand, E. Cocheret de la Morinière, and G. van der Velde. 2001. Dependence of Caribbean reef fishes on mangroves and seagrass beds as nursery habitats: a comparison of fish faunas between bays with and without mangroves/seagrass beds. Marine Ecology Progress Series 214:225–235.

Nagelkerken, I., M. Sheaves, R. Baker, and R. M. Connolly. 2015. The seascape nursery: a novel spatial approach to identify and manage nurseries for coastal marine fauna. Fish and Fisheries 16:362–371.

Pita, P., D. Fernádez-Márquez, and J. Freire. 2014. Short-term performance of three underwater sampling techniques for assessing differences in the absolute abundances and in the inventories of the coastal fish communities of the northeast Atlantic Ocean. Marine and Freshwater Research 65:105–113.

Rankey, E. C., and S. L. Reeder. 2012. Tidal sands of the Bahamian archipelago. Pages 537–566 *in* R. A. Davis, Jr. and R. W. Dalrymple, editors. Principles of tidal sedimentology. Springer, New York.

Sheaves, M. 2005. Nature and consequences of biological connectivity in mangrove systems. Marine Ecology Progress Series 302:293–305.

Smith, G. W. 1996. The Klebsiella-Halodule symbiosis: distribution and physiology of the endophyte. Pages 4–9 *in* N. B. Elliot, D. C. Edwards, and P. J. Godfrey, editors. Proceedings of the sixth symposium on the natural history of the Bahamas. Bahamian Field Station, San Salvador, Bahamas.

Sumner, D. M., and G. Belaineh. 2005. Evaporation, precipitation, and associated salinity changes at a humid, subtropical estuary. Estuaries 28:844–855.

Tzeng, W. N., and Y. T. Wang. 1992. Structure, composition and seasonal dynamics of the larval and juvenile fish community in the mangrove estuary of Tanshui River, Taiwan. Marine Biology 113:481–490.

UNEP (United Nations Environment Programme). 2006. Marine and coastal ecosystems and human well being: a synthesis report based on the findings of the millennium ecosystem assessment. UNEP, Nairobi, Kenya.

Walters, B. B., P. Rönnbäck, J. M. Kovacs, B. Crona, S. A. Hussain, R. Badola, J. H. Primavera, E. Barbier, and F. Dahdouh-Guebas. 2008. Ethnobiology, socio-economics and management of mangrove forests: a review. Aquatic Botany 89:220–236.

Ward-Paige, C., J. Mills Flemming, and H. K. Lotze. 2010. Overestimating fish counts by non-instantaneous visual censuses: consequences for population and community descriptions. PLOS ONE 5(9):e11722.

Willis, T. J. 2001. Visual census methods underestimate density and diversity of cryptic reef fishes. Journal of Fish Biology 59:1408–1411.

Wentworth, C. K. 1922. A scale of grades and class terms for clastic sediments. The Journal of Geology 30:377–392.

American Fisheries Society Symposium 83:81–83, 2015

EXTENDED ABSTRACT

The Conservation Implications of Spatial and Tidal Variability in the Use of Bahamian Tidal Mangrove Creeks by Transient Predatory Fishes

Brendan Talwar*
Shark Research and Conservation Program, Cape Eleuthera Institute
Post Office Box EL-26029, Rock Sound, Eleuthera, The Bahamas
and
Florida State University Coastal and Marine Laboratory
Florida State University
3618 Coastal Highway 98, St. Teresa, Florida, 32358, USA

Alastair R. Harborne
Marine Spatial Ecology Laboratory and
Australian Research Council Centre of Excellence for Coral Reef Studies
School of Biological Sciences, The University of Queensland
Goddard Building, Brisbane, QLD 4072, Australia

Edward J. Brooks
Shark Research and Conservation Program, Cape Eleuthera Institute
Post Office Box EL-26029, Rock Sound, Eleuthera, The Bahamas

Introduction

Caribbean mangroves, along with adjacent soft-bottom habitats, are recognized as important nurseries for a range of ecologically and commercially important fish species (e.g., Mumby et al. 2004). Research on the nursery function of tidal mangrove creeks, including their role as a refuge from predators, has tended to focus on juvenile fishes and their subsequent migration into adult habitats. In contrast, there are limited data on the transient predators of these juvenile fishes, meaning those individuals that move into shallow mangrove areas on rising tides to feed but leave there at low tide (Valentine-Rose and Layman 2011). This lack of data is surprising given the importance of these fishes.

This study aimed to provide insights into the identity of these diurnal, transient predators using mangrove creeks and how their use of this habitat is affected by water depth, flow rate, tidal range, and time of day. By studying three adjacent creeks in the same biophysical environment, this study also aimed to assess the degree of intrahabitat variability in creek use by these fishes.

Methods

Data were collected from three tidal creeks (Page, Kemps, and Broad) within 4 km at Cape Eleuthera, The Bahamas in September–October 2012. Creeks vary in their geomorphology and size of creek mouth. GoPro HD

* Corresponding author: talwar.brendan@gmail.com

Hero 3 underwater video cameras attached to cinderblocks were placed centrally in each creek mouth and began recording within 3 h of low tide to document fish entering the creeks as they flooded, with one camera deployed for every ~25 m of creek mouth. For each camera deployment, individual fish counts were combined to calculate the number of fish entering each creek during each hour surveyed after low tide (subsequently "time segment") and standardized as number of fish per minute.

These quantitative data were used in a permutational multivariate analysis of variance to assess whether there was any evidence of the entire transient predator assemblage varying among creeks or being significantly influenced by any abiotic factors. Explanatory continuous variables were hours since low tide, minutes since sunrise, a random variable representing camera deployment, the published low-tide depth during each tidal cycle, and a quadratic term for the number of hours since low tide (proxy for tidal flow). The categorical variable of creek identity was also included.

Univariate, species-specific analyses were then conducted. Data were transformed from number of fish seen per minute to presence/absence per time segment and analyzed using minimal adequate generalized linear models with binomial error structures and the logit link function using the same explanatory variables as described previously. Removal of nonsignificant terms led to a significant increase in model deviance and Akaike information criterion. Values were considered significant at $\alpha \leq 0.05$.

Results

A total of 13 diurnal camera deployments were conducted representing 55 one-hour segments. Video footage recorded four sufficiently abundant (seen in >25% of time segments) transient creek predators for detailed analyses: Bar Jack *Caranx ruber*, Lemon Shark *Negaprion brevirostris*, Houndfish *Tylosurus crocodilus*, and Great Barracuda *Sphyraena barracuda*.

None of the fixed factors were significant ($P > 0.075$) in the PERMANOVA model, indicating that the entire transient predator assemblage neither varied systematically among creeks nor was responding to any of the explanatory variables. The timings of the visits by all four species were explained by at least one of the abiotic variables, and the frequency of visits by two of the species also varied among creeks. Bar Jacks were seen significantly more frequent in Kemps Creek than Broad Creek and were absent from Page Creek. This species was also seen increasingly more frequently as the tidal cycle approached high tide. All but one Lemon Shark were juveniles and were more abundant in Page Creek than Broad Creek, with no individuals recorded in the mouth of Kemps Creek. Lemon Sharks were seen more frequently later in the day and were slightly more abundant during tidal cycles with a shallower low tide. Both Great Barracuda and Houndfish were seen more frequently just after low tide.

Discussion

Within the transient predatory fish assemblage, four species were particularly common, and their use of the creeks demonstrated that there is no single abiotic driver of habitat use as reflected in the multivariate analysis. This result extends earlier findings of spatially and temporally variable use of intertidal mangroves by resident and juvenile fishes (Ellis and Bell 2008).

It seems likely that mangrove creeks represent both important feeding and refuge areas for these species, although these data cannot point to a primary reason for their use. Water depth appears to be the main abiotic driver of the behavior of both Great Barracuda and Houndfish. Bar Jacks entered the creeks more frequently as high tide approached, suggesting that creek use is restricted to periods around high tide when the maximum area of wetlands is available. This addition of predatory species may contribute to the reduced abundance of some juvenile fishes around high tide compared to flood or ebb tides (Ellis and Bell 2008) due to increased predation risk (Lima and Dill 1990).

While there is certainly a need for additional research, the range of transient predators recorded during this study clearly highlights the importance of unique mangrove creeks, as well as the variability in their use. These findings underscore the need to conserve mangrove habitats and suggest that management schemes must target functionally distinct creeks rather than just including a representative creek within a protected area. Furthermore, the high relative abundance of Lemon Sharks in the relatively small Page Creek demonstrates that size is a poor proxy of functional importance and provides an example of the challenges faced when identifying critical sites for conservation using only remotely sensed imagery. Last, this study demonstrates the importance of water depth and access to shallow wetlands during high tides to transient predators, suggesting that restoring hydrological connectivity within mangrove systems can benefit these species by reestablishing access to forage and refuge habitats.

References

Ellis, W. L., and S. S. Bell. 2008. Tidal influence on a fringing mangrove intertidal fish community as observed by in situ video recording: implications for studies of tidally migrating nekton. Marine Ecology Progress Series 370:207–219.

Lima, S. L., and L. M. Dill. 1990. Behavioral decisions made under the risk of predation: a review and prospectus. Canadian Journal of Zoology 68:619–640.

Mumby, P. J., A. J. Edwards, J. E. Arias-González, K. C. Lindeman, P. G. Blackwell, P. G., A. Gall, M. I. Gorczynska, A. R. Harborne, C. L. Pescod, H. Renken, C. C. C. Wabnitz, and G. Llewellyn. 2004. Mangroves enhance the biomass of coral reef fish communities in the Caribbean. Nature 427:533–536.

Valentine-Rose, L., and C. A. Layman. 2011. Response of fish assemblage structure and function following restoration of two small Bahamian tidal creeks. Restoration Ecology 19:205–215.

Ecological Services of Mangroves

American Fisheries Society Symposium 83:87–99, 2015

Secondary Productivity of Mollusks in Mangrove Estuaries of Ensenada de La Paz, Baja California Sur, Mexico

ESTEBAN F. FÉLIX-PICO* AND MAURICIO RAMÍREZ-RODRÍGUEZ
Centro Interdisciplinario de Ciencias Marinas
Instituto Politécnico Nacional
Avenida Instituto Politécnico Nacional s/n
Colonia Playa Palo de Santa Rita, La Paz, Baja California Sur, 23096, México

JORGE A. LÓPEZ-ROCHA
Unidad Multidisciplinaria de Docencia e Investigación, Facultad de Ciencias
Universidad Nacional Autónoma de México, Sisal, Yucatán, C.P. 97130, México

Abstract.—To measure secondary productivity of mangrove systems, we estimated the abundance (individuals/m^2) and mass increment (g/month) of the two bivalve species: the black ark *Anadara tuberculosa* and palmate oyster *Saccostrea palmula*. Mass increments were based on individual growth rates derived from length-frequency distributions analyses. Samples were collected at three mangrove estuaries in a sand barrier at Ensenada de La Paz from August 2007 to July 2009. The average abundance was 1.27 individuals/m^2 for black ark and 510 individuals/m^2 for palmate oysters. Estimated growth rates were 3.67 g/month for black ark and 0.18 g/month for palmate oysters. The average secondary productivity of the black ark was 4.51 g·m^{-2}·month^{-1} and peaked during the spring, while for the palmate oyster *Saccostrea palmula* it was 97.9 g·m^{-2}·month^{-1}, with peak productivity recorded during the summer. The findings of this investigation constitute a necessary element for establishing a baseline to evaluate the consequences of the various natural and anthropogenic pressures that the mangrove systems of El Mogote of La Paz Bay, Baja California Sur, Mexico.

Introduction

Mangrove ecosystems are critical habitats for a large variety of mollusks, crustaceans, and fish of economic importance and ecological interest. The mangroves create unique ecological environments that host rich assemblages of species, which makes these zones one of the most productive ecosystems (Mann 2000; Aburto-Oropeza et al. 2008). Mangroves are also important because they provide food and shelter for a large population of juvenile organisms of many species.

Mangrove productivity is fundamental because of their contribution of organic detritus and secondary productivity. Mangrove systems are considered one of the most productive ecosystems on earth; they are four times as productive as the continental shelf and 40 times as productive as oceanic regions (Berger et al. 1989).

Mollusks inhabiting mangrove ecosystems are mostly gastropods and bivalves, which are commonly found attached to the

* Corresponding author: efelix@ipn.mx

roots of mangrove trees; examples include *Littorina* and *Nerita* snails and *Crassostrea* and *Saccostrea* oysters (Kathiresan and Binghan 2001; Prieto et al. 2008). In muddy substrates, the bivalves *Chione* spp., *Tagelus* spp., *Anadara* spp., and *Mytella* spp. are common (Lacerda et al. 2002). In the mangrove systems fringing La Paz Bay, two of the most common bivalve species are the black ark *Anadara tuberculosa* and the palmate oyster *Saccostrea palmula*.

The black ark is a species of great economic importance and is harvested throughout the Pacific littoral (MacKenzie, 2001). This species is associated with the red mangrove *Rhizophora mangle* and the white mangrove *Laguncularia racemosa* in the intertidal zone, where it lives buried in the mud. It is distributed in the eastern Pacific, from the coasts of the Mexican state of Baja California Sur (B.C.S.) south to Peru (Keen 1971; Coan and Valentich-Scott 2012). In Mexico, it is mainly harvested in Magdalena Bay, Pacific coast of B.C.S. (Hernández-Valenzuela 1996; Félix-Pico et al. 2009).

On the other hand, the mangrove palmate oyster is attached to the roots of the red and white mangroves and has little commercial value. It is distributed from San Ignacio Lagoon in the Mexican state of Baja California to Panama. It is also found in the Galapagos Islands of Ecuador and Cocos Island, Costa Rica. (Keen 1971; Cruz and Jiménez 1994; Coan and Valentich-Scott 2012).

Mangrove areas have been seriously reduced in recent decades due to human activities like sewage disposal, constructions of highways and piers, aquaculture, fishery, and tourism (Valiela et al. 2001). For this reason, the objective of this investigation was to determine biomass increments of black ark and palmate oyster as indicators of secondary productivity in mangrove systems in Ensenada de La Paz, B.C.S., Mexico, to set out a baseline that will allow measurement of anomalies due to potential natural or anthropogenic perturbations.

Study Site

The study was developed at three hypersaline (>36 practical salinity units) and warm (>21°C) estuaries, part of a small mangrove ecosystem located at the sandbar that separates La Paz Bay and the Ensenada de La Paz (Figure 1). The Ensenada is 50-km^2 lagoon (24°08'N, 110°23'W). The separation between the lagoon and the bay is a sand spit or bar called El Mogote.

Methods

To estimate secondary productivity in El Mogote estuary, the two most abundant bivalve species were selected: the black ark and the palmate oyster. Three estuaries were chosen with a high coverage of white mangrove and red mangrove (Figure 1).

Within each estuary, sampling was carried out monthly from August 2007 through July 2008 and once every 3 months from October 2008 through June 2009. All black arks observed during ca. 30-min searches through the estuaries were collected. The area covered by these searches was estimated to the nearest square meter by multiplying the transect length by the 2-m-wide search area.

As for oysters, we noticed that these were particularly abundant at the mouth of estuary A; hence, it was decided that both the inner (A1) and mouth (A2) zones of the estuary would be sampled. At each sampling site, four red mangrove roots were cut and all specimens present were counted. Abundance per square meter was estimated and specimens collected were measured (shell length, height, and width) and weighed.

The soft parts were removed and drained to dry it (10 min on a sieve). Both soft part and empty shells were weighed. To obtain dry weight of the tissues, the separated individuals were placed into labeled aluminum foil cages and dried in an oven at 70–80°C to constant weight (about 72 h). Dry parts were weighed using a balance with an accuracy of 0.001 g.

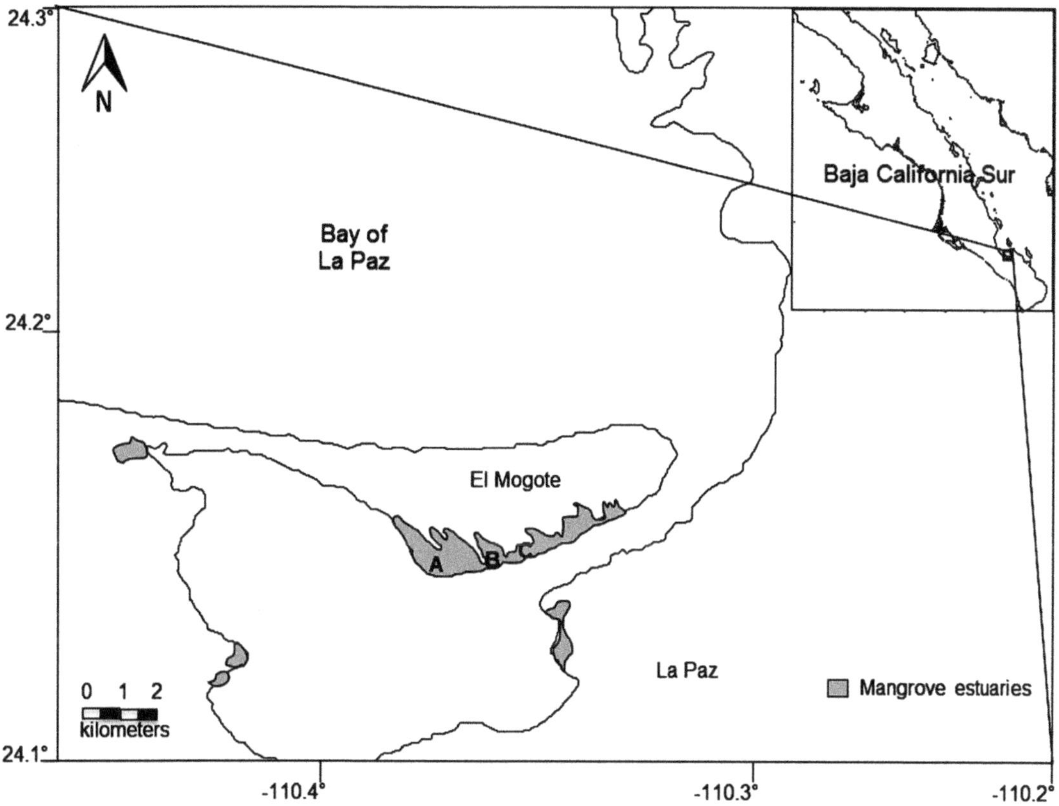

Figure 1.—Study area in El Mogote, La Paz Bay, Baja California Sur, Mexico. The locations of the three estuaries (A, B, and C) where secondary productivity was estimated are shown.

The estimation of secondary productivity in mangrove bivalves was based on biomass increase. Living tissue production for each species was estimated as

$$P = N \cdot B,$$

where P is productivity, N is the density during the sampling period, and B is the increase in average mass over the same period of time.

Mass increment (B) was estimated by determining the rate of individual growth for each species using size-frequency distribution analysis. Measurements of total shell length were grouped into monthly length-frequency distributions with 2-mm-wide intervals. The Battacharya method was used to identify modal length-frequency distribution groups, which were assumed to represent cohorts or age-groups (Sparre and Venema 1995). Average and standard deviation of length were calculated for each modal group identified.

Modal groups were tracked over time to measure length increase due to the individuals' body growth. These methods were implemented using FISAT II (Gayanilo et al. 2004) analysis software. Based on this information, growth rates were calculated in millimeters per month. To estimate individual mass increment in grams per month, the relationship between shell-free mass (W) and shell length (L) was described as follows:

$$L = a \cdot W^{b},$$

where a and b are model parameters (Sparre and Venema 1995).

In addition, recruitment patterns for each species were estimated using the methods described by Moreau and Cuende (1991) and Pauly (1983). Differences in secondary productivity between estuaries were tested by means of the Kruskal-Wallis rank-sum test, since the data did not follow a normal distribution.

Results

Black ark

A total of 2,412 specimens with an average total length of 65.0 mm (range 32.8–97.5 mm) were collected. Total mass ranged from 7.5 to 275.0 g with an average of 86.7 g, and the shell-free mass ranged from 2.1 to 50.8 g with an average of 16.0 g. Estimated dry weight ranged between 0.1 and 7.6 g, with an average of 3.4 g.

Average density by estuary was 1.76 ± 0.74 (SD), 1.28 ± 0.49 (SD), and 0.60 ± 0.46 (SD) individuals per square meter (individuals/m^2) for estuaries A, B, and C, respectively, with an overall average of 1.27 ± 0.75 (SD) individuals/m^2. Minimum and maximum values were 0.15 individuals/m^2 (estuary C, October 2007) and 3.45 individuals/m^2 (estuary A, April 2009; see Table 1). Between August 2007 and July 2008, several modal groups were identified (Figure 2); based on this information, length increase over time was estimated (Figure 3). The estimated growth rate ranged from 3.81 to 5.51 mm/month, with an average of 4.91 ± 0.64 (SD).

The relationship between shell-free mass (W) and shell length (L) was represented by the equation: $W = 0.00006 \cdot L^{2.97}$. From this, mass gain rates were calculated, ranging from 2.64 to 5.47 g/month, with an average of 3.67 g/month ± 1.06 (SD). Biomass production due to body growth was, on average, 4.51 g·m^{-2}·month^{-1} ± 2.62 (SD). The

Table 1.—Estimates of abundance and biomass production due to body growth in black ark in El Mogote, La Paz Bay, Baja California Sur (CV = coefficient of variation; Ind. = individuals.)

	Estuary A		Estuary B		Estuary C	
Date	Ind. per m^2	Production g·m^{-2}·month^{-1}	Ind. per m^2	Production g·m^{-2}·month^{-1}	Ind. per m^2	Production g·m^{-2}·month^{-1}
Aug 7	2.15		0.64	2.27		
Sept 7	1.49	5.29	0.52	1.85		
Oct 7	2.02	7.19	1.93	6.87	0.15	0.53
Nov 7	1.49	5.29	1.30	4.62		
Dec 7	2.18	7.75	1.07	3.79	0.30	1.07
Jan 8	1.43	5.08	1.71	6.09	0.43	1.54
Feb 8	1.15	4.09	1.91	6.80	1.00	3.55
Mar 8	3.00	10.65	1.18	4.19	0.97	3.44
Apr 8	2.28	8.10	1.00	3.56	1.74	6.20
May 8	1.53	5.44	1.07	3.79	0.16	0.57
Jun 8	1.43	5.08	1.71	6.09	0.43	1.54
Jul 8	0.59	2.08	0.72	2.56	0.33	1.19
Oct 8	1.77	6.30	0.78	2.76	0.43	1.52
Jan 9	1.29	4.58	2.09	7.41	0.83	2.96
Apr 9	3.45	12.26	1.50	5.33	0.42	1.51
Jul 9	0.93	3.32	1.36	4.83		
Average	1.76	6.17	1.28	4.55	0.60	2.14
SD	0.74	2.68	0.49	1.75	0.46	1.64
CV	42%	43%	38%	38%	77%	77%

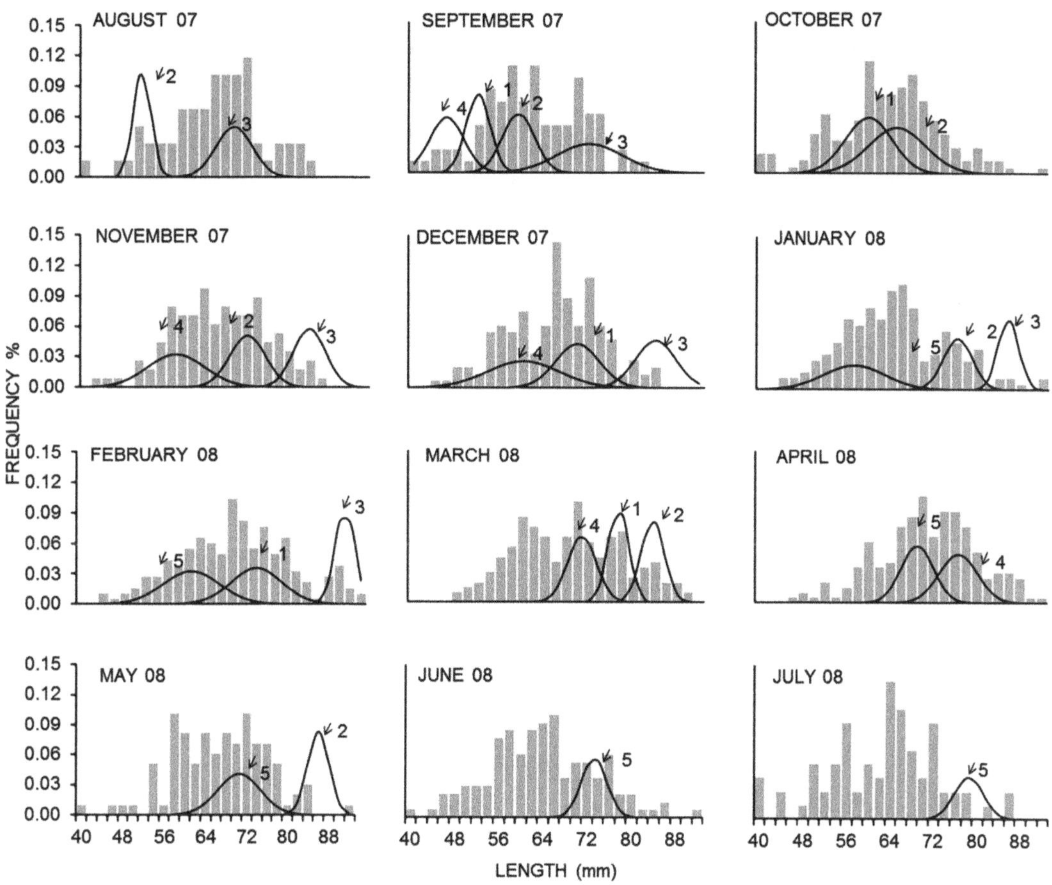

Figure 2.—Length-frequency distributions of the black ark in El Mogote, La Paz Bay, Baja California Sur. The numbers indicate the modal groups identified and their monthly progression.

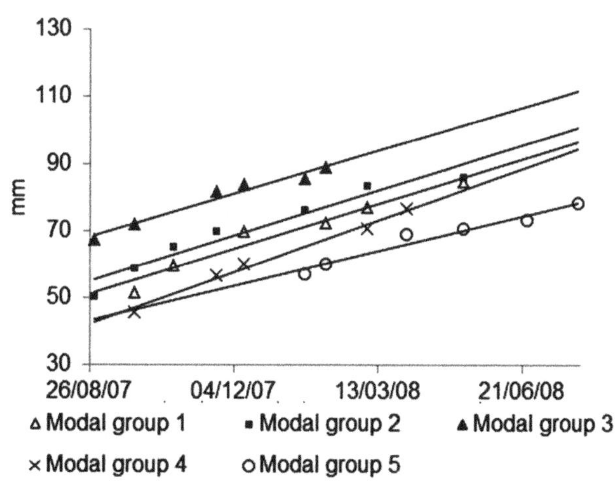

Figure 3.—Average length increase for each of the modal groups identified in the length-frequency distributions of the black ark in El Mogote, La Paz Bay, Baja California Sur.

highest productivity was recorded in estuary A (6.17 g·m^{-2}·month^{-1}, ±2.68 SD), followed by estuary B (4.55 g·m^{-2}·month^{-1}, ±1.75 SD) and, then, by estuary C (2.14 g·m^{-2}·month^{-1}, ±1.64 SD, Table 1). Maximum productivity occurred during the spring in estuaries A and C (Figure 4a, 4c) while peak productivity in estuary B was recorded during the winter (Figure 4b). In 2007–2008 and 2008–2009, the same temporal pattern of productivity was observed in estuaries A and B, with particularly high productivity in 2008–2009. In estuary C, during 2008–2009, the highest productivity was recorded in the winter, showing a shift from 2007 to 2008, when the highest productivity occurred in the spring (Figure 4b).

Palmate oyster

A total of 1,068 oysters were collected, with an average total length of 31.1 mm (range

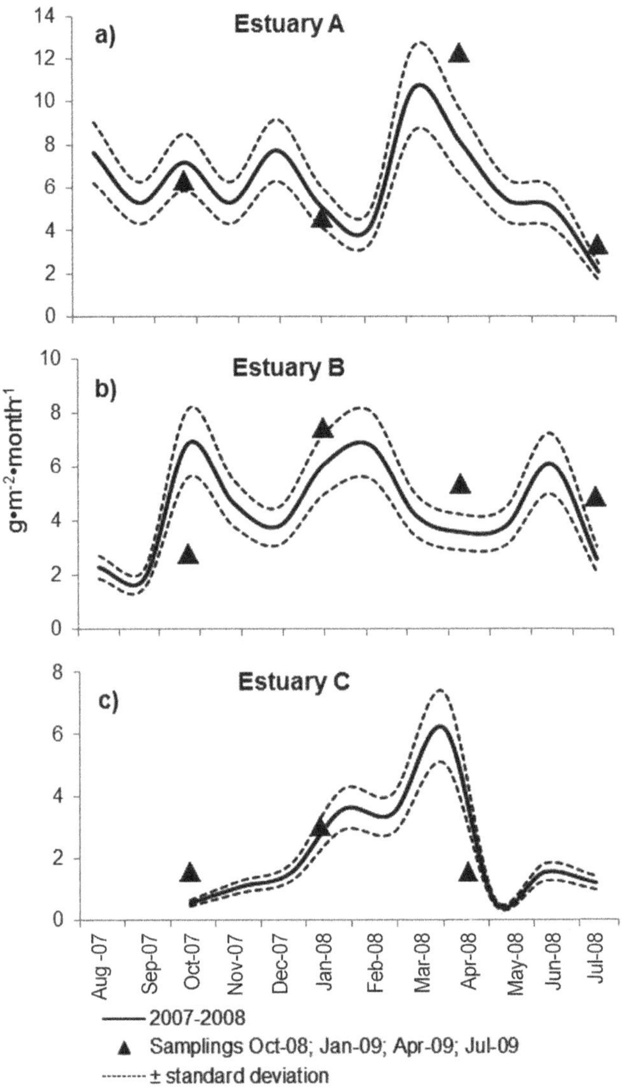

Figure 4.—Temporal variation in secondary productivity due to body growth of the black ark in El Mogote, La Paz Bay, Baja California Sur.

11.0–66.2 mm). Total mass ranged from 4.1 to 76.4 g, with an average of 19.7 g. The shell-free mass ranged from 0.5 to 6.0 g, with an average of 2.3 g, and the dry weight ranged from 0.11 to 1.25 g, averaging 0.46 g.

Estimated population density was 510 individuals/m^2 ± 333 (SD), with a minimum of 80 individuals/m^2 in estuary A1 in January 2009 and a maximum of 1,720 individuals/m^2 in estuary A1 in June 2007. The average density by estuary was 413 ± 409 (SD), 573 ± 360 (SD), 395 ± 136 (SD), and 661 ± 312 (SD) individuals/m^2 for estuaries A1, A2, B, and C, respectively (Table 2).

From August 2007 to July 2008, several modal groups were identified (Figure 5) from which growth rates were estimated. Growth rates varied between 2.8 and 5.1 mm/month, with an average of 3.8 mm/month ±1.2 (SD, Figure 6).

The relationship between shell-free mass (W) and shell length (L) was described by the equation $W = 0.012 \cdot L^{1.31}$. From it, the estimated rate of mass gain due to body growth varied from 0.12 to 0.25 g/month, with an average of 0.18 g/month ± 0.05 (SD).

The increase in biomass due to body growth was, on average, 97.9 g·m^{-2}·month^{-1}, ±63.9 (SD). The peak productivity was recorded in estuary C (127.0 g·m^{-2}·month^{-1}, ±59.9 SD) followed by estuary A2 (110.0 g·m^{-2}·month^{-1}, ±69.3 SD), inlet A1 (79.4 g·m^{-2}·month^{-1}, ±78.5 SD), and then estuary B (76.0 g·m^{-2}·month^{-1}, ±26.3 SD, Table 2). Productivity in estuary C was significantly higher than in estuaries A1 and B ($p < 0.05$). The maximum productivity was observed in October 2008 and July 2009 in estuaries A1, A2, and B (Figure 7a–c), while in estuary C productivity remained constant (Figure 7d). Variations in biomass production in the 2007–2008 and 2008–2009 periods showed similar temporal productivity patterns in all four estuaries (Figure 7a–d).

Recruitment occurred 11 months out of the year. There was no recruitment for palmate oysters in March and none for black ark in December. The main season for black ark was from June through September (Figure 8a), while for palmate oyster occurred mainly from September through December (Figure 8b).

Discussion

The estimated population densities of the two species were highly variable, which is likely related to local environmental conditions and survival and recruitment rates. As a response to environmental conditions, these bivalves show different preferences for and tolerances to temperature, salinity, substrate composition, and tidal levels. The largely different densities of these two species (1.27 individuals/m^2 for black ark and 510 individuals/m^2 for palmate oyster) was due to the three-dimensional use of space by palmate oysters (from 1.0 to 1.5 m) attached to mangrove roots.

The estimated densities of black ark are low compared to those found in Magdalena Bay, B.C.S., Mexico, with an average of 12 individuals/m^2 (Baqueiro 1980), and for eared ark Anadara notabilis (Röding 1798) in Venezuela, with 14 individuals/m^2 (Prieto et al. 1998).

However, densities of black ark found in El Mogote estuaries are within the range found in Colombia and Costa Rica (0.5–4.23 individuals/m^2, Squires et al. 1975; Campos et al. 1990; Borda and Cruz 2004). The high density of black ark recorded in Magdalena Bay, B. C. S., makes this the most important region for black ark harvest in Mexico, in contrast to El Mogote, where no fishery for this species has developed.

The densities of palmate oyster found in this work were higher than those reported for other bivalves in a range of environments and are comparable only to those found for mangrove oyster *Crassostrea rhizophorae* (818 individuals/m^2; Prieto et al. 2008).

Length-frequency distributions for both species were multimodal, which is consistent with the continuous or prolonged re-

Table 2. Estimates of abundance and biomass production due to body growth in palmate oyster in El Mogote, La Paz Bay, B.aja California Sur (CV = coefficient of variation; Ind. = individuals)

	Estuary A1		Estuary A2		Estuary B		Estuary C	
Date	Ind. per m^2	Production $g \cdot m^{-2} \cdot month^{-1}$	Ind. per m^2	Production $g \cdot m^{-2} \cdot month^{-1}$	Ind. per m^2	Production $g \cdot m^{-2} \cdot month^{-1}$	Ind. per m^2	Production $g \cdot m^{-2} \cdot month^{-1}$
Jul 7	1,720	330.24			180	34.56	320	61.44
Aug 7	670	128.64	1,050	201.60	510	97.92	780	149.76
Sep 7	340	65.28	1,030	197.76	460	88.32	600	115.20
Oct 7	120	23.04	230	44.16	120	23.04	730	140.16
Nov 7	810	155.52	230	44.16	510	97.92	530	101.76
Dec 7	220	42.24	490	94.08	380	72.96	810	155.52
Jan 8	220	42.24	490	94.08	380	72.96	810	155.52
Feb 8	220	42.24	490	94.08	380	72.96	810	155.52
Mar 8	220	42.24	490	94.08	380	72.96	810	155.52
Apr 8	220	42.24	490	94.08	380	72.96	810	155.52
May 8	540	103.68	330	63.36	420	80.64	250	48.00
Jun 8	220	42.24	490	94.08	380	72.96	810	155.52
Jul 8	190	36.48	630	120.96	730	140.16	1,440	276.48
Oct 8	800	153.60	1,540	295.68	370	71.04	110	21.12
Jan 9	80	15.36	710	136.32	520	99.84	310	59.52
Apr 9	340	65.28	270	51.84	270	51.84	840	161.28
Jul 9	100	19.20	210	40.32	360	69.12	480	92.16
Average	413.53	79.40	573.13	110.04	395.88	76.01	661.76	127.06
SD	408.99	78.53	360.81	69.28	136.89	26.28	311.91	59.89
CV	99%	99%	63%	63%	35%	35%	47%	47%

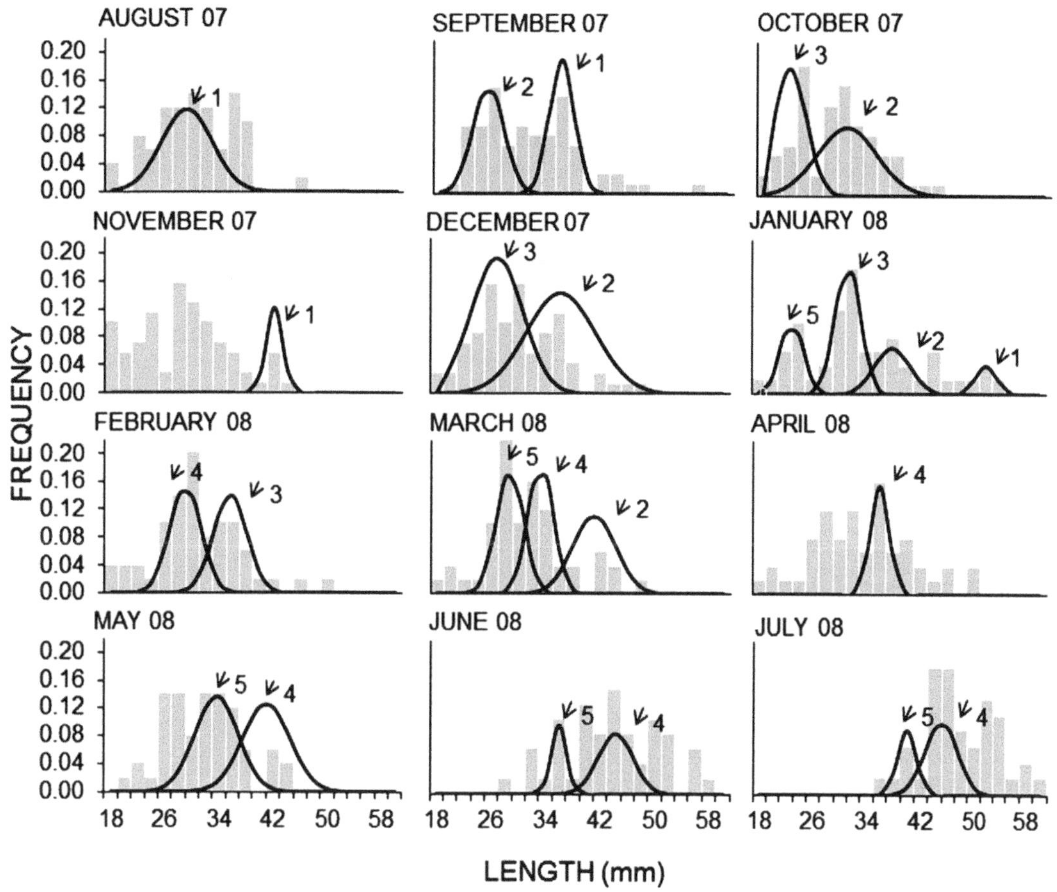

Figure 5.—Length-frequency distributions of the palmate oyster in El Mogote, La Paz Bay, Baja California Sur. The numbers indicate the modal groups identified and their monthly progression.

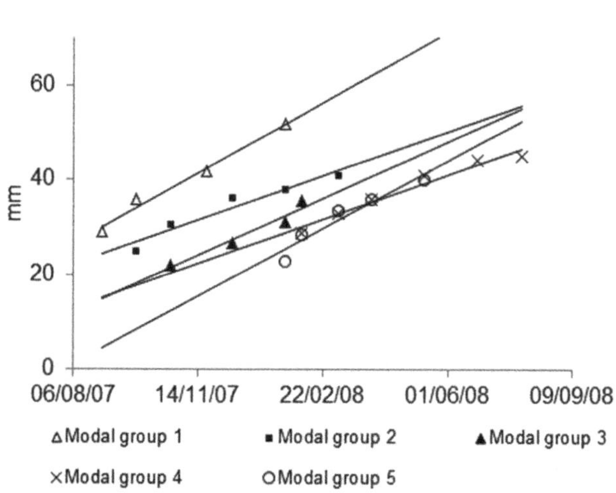

Figure 6.—Average size increase for each of the modal groups identified in the length-frequency distributions of the palmate oyster in El Mogote, La Paz Bay, Baja California Sur.

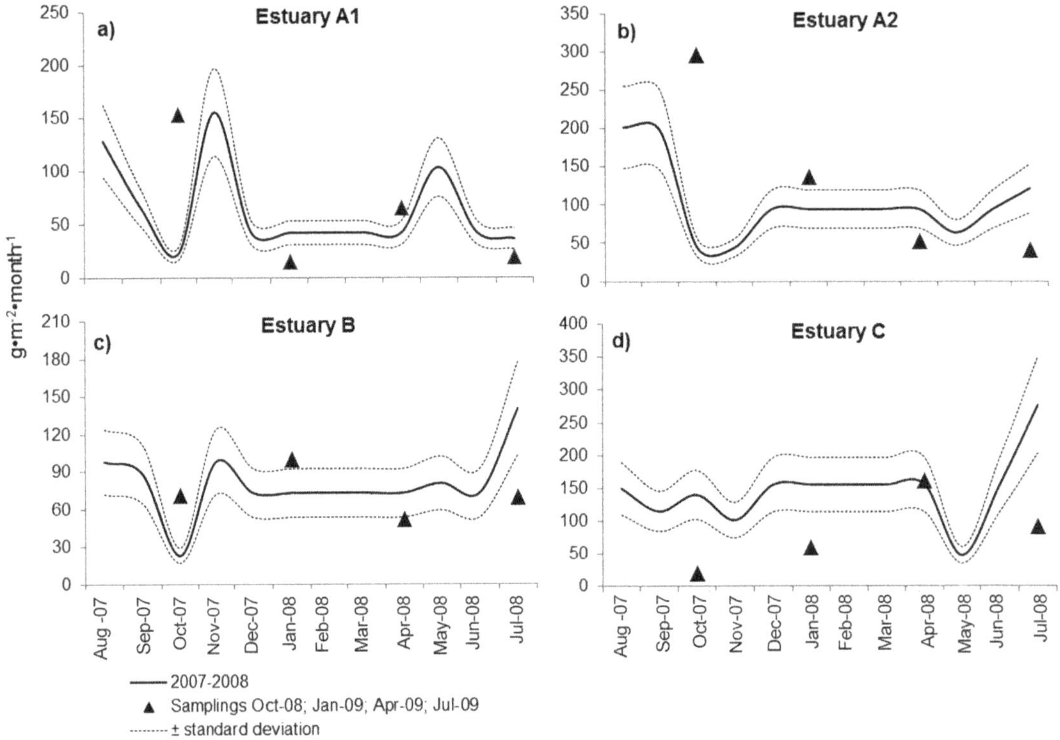

Figure 7.—Temporal variation in secondary productivity due to body growth of the palmate oyster in El Mogote, La Paz Bay, Baja California Sur.

cruitment and reproduction patterns reported for these species (Pérez Medina 2005; Romo-Piñera 2005; Félix-Pico et al. 2007). Within each length-frequency distribution, five modal groups were identified in which monthly increases in mean length were observed. This growth estimate for black ark was deemed appropriate as the increase in average length was similar in four of the five modal groups. However, the palmate oyster shows high variability that is not clear now.

It was not possible to compare secondary productivity of these bivalves in the same zone since there were no previous studies on this subject. However, mean secondary productivity estimated for the black ark in this study (4.51 g·m^{-2}·month^{-1}) lies between 0.8 and 16.3 g·m^{-2}·month^{-1} reported for other bivalves (false tulip mussel *Modiolus squamosus*, trigonal tivela *Tivela mactroides*, cross-barred venus *Chione cancellata*, Atlantic turkey wing *Arca zebra*, and purplish scallop *Argopecten purpuratus* (Prieto et al. 1985, 1998, 2001; Tata and Prieto 1991; Tarazona et al. 2007). It is difficult to ascertain if these comparisons are appropriate due to lack of knowledge on growth characteristics of these species. The estimated productivity of palmate oyster (97.9 g·m^{-2}·month^{-1}) is less than the 196.3 g·m^{-2}·month^{-1} estimated for mangrove oyster (Prieto et al. 2008).

The between-estuary variation in secondary productivity is directly proportional to the density values found. The lowest secondary productivity for black ark, recorded in the summer (Figure 4a–c, Table 1), was related to a higher recruitment in those months (June–September, Figure 8a), when there is a larger proportion of small individuals in the

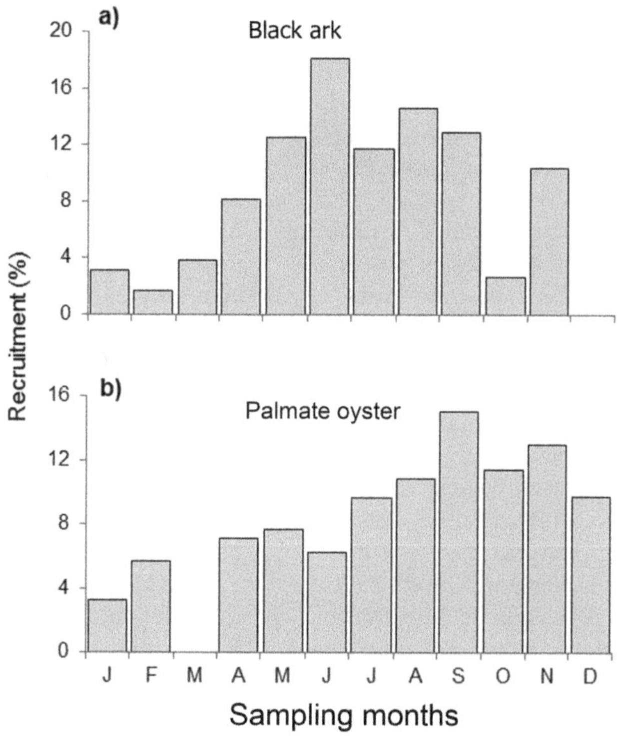

Figure 8.—Recruitment patterns in the black ark and palmate oyster in El Mogote, La Paz Bay, Baja California Sur.

population, a pattern that is similar to, although not as clear as, that of the mangrove palmate oyster (Figure 8b).

Figures 4a–c and 7a–d show the comparison of monthly estimates of secondary productivity for the first year of the study (2007–2008) and quarterly estimates for the following year (2008–2009). Similar variation patterns are observed throughout, which suggests a period of stability both in terms of environmental conditions and in possible effects of human activities such as pollution, habitat degradation, and harvest.

Given the significance of mangroves as critical habitat for numerous mollusk, crustacean, fish, bird, and mammal species (Lara-Domínguez and Yáñez-Arancibia 1999), the conservation and appropriate management of mangrove ecosystems is of utmost importance. The findings of this investigation constitute a necessary element for establishing a baseline to evaluate the consequences of the various natural and anthropogenic pressures that the mangrove systems of El Mogote of La Paz Bay, B.C.S., Mexico.

Acknowledgments

The authors thank Paraíso del Mar Resort for funding this project, with additional support from the Instituto Politécnico Nacional of Mexico through SIP and COFAA. Maria Elena Sánchez and Lynna Kiere contributed to the English version of the manuscript.

References

Aburto-Oropeza, O., E. Ezcurra, G. Danemann, V. Valdez, J. Murray, and E. Sala. 2008. Mangroves in the Gulf of California increase fishery yields. The National Academy of Sciences of the United States of America 105:10456–10459.

Baqueiro, E. 1980. Population structure of the mangrove cockle *Anadara tuberculosa* (Sowerby, 1833) from eight mangroves swamps in Magdalena and Almejas bays, Baja California Sur, Mexico. Proceedings of the National Shellfisheries Association 70:201–206.

Berger, W. H., V. S. Smetacek, and J. S. Gray. 1989. Productivity of the ocean: present and past. Wiley-Interscience, New York.

Borda, C. A., and R. Cruz. 2004. Pesca artesanal de bivalvos (*Anadara tuberculosa* y *A. similis*) y su relación con eventos ambientales. Pacífico colombiano. [Artisanal fishery of bivalves (*Anadara tuberculosa* and *A. similis*) and its relationship with environmental events in the Columbian Pacific.] Revista de Investigaciones Marinas 25:197–208.

Campos, J. A., M.-L. Fournier, and R. Soto. 1990. Estimación de la población de *Anadara tuberculosa* (Bivalvia: Arcidae) en Sierpe Térraba, Costa Rica. [Population assessment of *Anadara tuberculosa* (Bivalvia: Arcidae) in Sierpe-Térraba, Costa Rica.] Revista de Biología Tropical 32:477–480.

Coan, E. V., and P. H. Valentich-Scott. 2012. Bivalves seashells of tropical West America: marine bivalve mollusks from Baja California to northern Peru. Santa Barbara Museum of Natural History, Monographs 6, Santa Barbara, California.

Cruz, R., and J. Jiménez. 1994. Moluscos asociados a las áreas de manglar de la Costa Pacífica de Americanérica Central. [Associated mollusk species to the mangrove areas of the Pacific coast of Central America.] Editorial Fundación UNA, Heredia, Costa Rica.

Félix-Pico, E. F., F. A. García-Domínguez, O. E. Holguín-Quiñones, and J. A. López-Rocha. 2007. Ciclo reproductivo del ostión de mangle, *Crassostrea palmula* (Bivalvia: Ostreidae) en el estero El Conchalito, La Paz, B. C. S. [Reproductive cycle of the mangrove oyster *Saccostrea palmula* (Bivalvia: Ostreidae) in the El Conchalito estuary, La Paz, South Baja California.] Pages 255–256 *in* E. Ríos-Jara, M. C. Esqueda-González, and C. M. Galván-Villa, editors. Estudios sobre la Malacología y Conquiología en México. [Studies of malacology and conchiology in Mexico.] Universidad de Guadalajara, Guadalajara, Jalisco, Mexico.

Félix-Pico, E. F., E. M. Ramírez-Rodríguez, and O. E. Holguín-Quiñones. 2009. Growth and fisheries of the black ark *Anadara tuberculosa*, a bivalve mollusk, in Bahía Magdalena, Baja California Sur, Mexico. North American Journal of Fisheries Management 29:231–236.

Gayanilo, F. C., P. Sparre, and D. Pauly. 2004. FAO-ICLARM stock assessment tools II Windows version (FISAT II) user's guide (revision I). Food and Agriculture Organization of the United Nations, FAO Computerized Information Series (Fisheries) No 8, Rome.

Hernández-Valenzuela, R. 1996. Pesquería de almeja pata de mula. [Fisheries of the black ark]. Pages 121–126 *in* M. Casas-Valdez, M. and G. Ponce-Díaz, editors. Estudio del potencial pesquero y acuícola de Baja California Sur, volume I. [Study of the potential fisheries and aquaculture of Baja California Sur, volume I.] Secretaría Medio Ambiente Recursos Naturales y Pesca México, Centro Interdisciplinario de Ciencias Marinas, Centro de Investigaciones Biológicas del Noroeste, Food and Agriculture Organization of the United Nations, México.

Kathiresan, K., and B. L. Bingham. 2001. Biology of mangroves and mangrove ecosystems. Advances in Marine Biology 40:81–252.

Keen, A. M. 1971. Sea shells of tropical West America: marine mollusks from Baja California to Peru, 2nd edition. Stanford University Press, Stanford, California.

Lacerda, L. D., J. E. Conde, B. Kjerfve, R. Alvarez-Leon, C. Alarcon, and J. Plania. 2002. American mangroves. Pages 1–62 *in* L. D. Lacerda, editor. Mangrove ecosystems: function and management. Springer, New York.

Lara-Domínguez, A. L., and A. Yáñez-Arancibia. 1999. Productividad secundaria, utilización del hábitat y estructura trófica. [Secondary production, use of habitat, and trophic structure.] Pages 153–166 *in* A. Yáñez-Arancibia and Lara-Domínguez, editors. Ecosistemas de manglar en Americanérica tropical. [Mangrove ecosystem in tropical America.] Instituto de Ecología, A.

C. México; UICN/ORMA, Costa Rica; and NOAA/NMFS, Silver Spring, Maryland.

MacKenzie, C. L. 2001. The fisheries for mangrove cockles, *Anadara* spp., from Mexico to Peru, with descriptions of their habitats and biology, the fishermen's lives, and the effects of shrimp farming. Marine Fisheries Review 63:1–39.

Mann, K. H. 2000. Ecology of coastal waters with implications for management. Blackwell Scientific Publications, Oxford, UK.

Moreau, J., and F. X. Cuende. 1991. On improving the resolution of the recruitment patterns of fishes. ICLARM Fishbyte 9:45–46.

Pauly, D. 1983. Some simple methods for the assessment of tropical fish stocks. FAO Fisheries Technical Paper 234:1–52.

Pérez Medina, D. R. 2005. Biología reproductiva de *Anadara tuberculosa* (Bivalvia: Arcidae) en el estero Santo Domingo. [Reproductive biology of *Andara tuberculosa* (Bivalvia: Arcidae) in the Santo Domingo estuary.] Tesis de Maestría, Centro Interdisciplinario de Ciencias Marinas, IPN, La Paz, B.C.S., Mexico.

Prieto, A., R. Pereira, and R. Manrique. 1985. Producción secundaria del mejillón *Modiolus squamosus* en Tocuchare, Golfo de Cariaco, Venezuela. [Secondary production of the mussel *Modiolus squamosus* in Tocuchare, Cariaco Gulf, Venezuela.] Acta Científica Venezolana 36:258–264.

Prieto, A., C. Ramos, and D. Arrieche. 1998. Producción secundaria de *Chione cancellata* (Bivalvia: Veneridae) de la costa sur del Golfo de Cariaco, Venezuela. [Secondary production of *Chiona cancellata* (Bivalvia: Veneridae) on the south coast of the Cariaco Gulf, Venezuela]. Revista de Biología Tropical 46:913–919.

Prieto, A., O. Ramos, D. Arrieche, J. Villalba, and C. Lodeiros. 2001. Producción secundaria e índice de condición en *Arca zebra* (Mollusca: Bivalvia) del Golfo de Cariaco, Venezuela. [Secondary production and condition index of the *Arca zebra* (Molluska: Bivalvia) from Cariaco Gulf, Venezuela.] Revista de Biología Tropical 49:599–608.

Prieto, A., A. Montes, and L. J. Ruiz. 2008. Potencial de producción de biomasa en una población natural de la ostra *Crassostrea rhizophorae*, en la laguna Grande de Obispo, Golfo de Cariaco, Venezuela. [Potential production of biomass in a natural production of the oyster *Crassostrea rhizophorae*, in the Grande de Obispo lagoon, Cariaco Gulf, Venezuela.] Interciencia 33:747–753.

Romo-Piñera, A. 2005. Ciclo reproductivo del ostión de mangle *Saccostrea palmula* (Carpenter 1857), en Bahía Magdalena, Baja California Sur. [Reproductive cycle of the mangrove oyster *Saccostrea palmula* (Carpenter 1857) in Magdalena Bay, South Baja California.] Tesis de Maestría, Centro Interdisciplinario de Ciencias Marinas, IPN, La Paz, B.C.S., México.

Sparre, P., and S. C. Venema. 1995. Introducción a la evaluación de recursos pesqueros tropicales. [Introduction to tropical fish stock assessment. Part I manual.] FAO Fisheries Technical Paper 306:1–96.

Squires, H., M. Estevez, O. Barona, and O. Mora. 1975. Mangrove cockles, *Anadara* spp. of the Pacific coast of Colombia. The Veliger 18:57–68.

Tarazona, J., R. Espinoza, M. Solís, and W. Arntz. 2007. Crecimiento y producción somática de la concha de abanico (*Argopecten purpuratus*) en Bahía Independencia, Pisco (Perú) comparados entre eventos El Niño y La Niña. [Growth and somatic production of the purplish scallop (*Argopecten purpuratus*) in Independence Bay, Pisco (Peru), comparing between El Niño and La Niña events.] Revista de Biología Marina y Oceanografía 42:275–285.

Tata, A., and A. Prieto. 1991. Producción secundaria en una población del bivalvo tropical *Tivela mactroides* (Veneridae) en el Oriente de Venezuela. [Secondary production of a population of the tropical bivalve *Tivela mactroides* (Veneridae) in western Venezuela.] Caribbean Journal of Science 21:63–73.

Valiela, I., J. L. Bowen, and J. K. York. 2001. Mangrove forests: one of the world's threatened major tropical environments. BioScience 51:807–815.

American Fisheries Society Symposium 83:101–112, 2015

Carbon Sequestration of Dwarf Red Mangrove in The Bahamas

Chelsea R. Barreto, Pedram P. Daneshgar*, and John A. Tiedemann
Marine and Environmental Biology and Policy Program, Monmouth University
400 Cedar Avenue, West Long Branch, New Jersey 07764, USA

Abstract.—Mangrove ecosystems are being lost globally at an alarming rate due to deforestation, reclamation, and urbanization. Not only is the loss of these ecosystems detrimental to the commercially and ecologically important marine species they support, there is also a reduction in the ecosystem services they provide, namely mitigating rising carbon dioxide levels by serving as carbon sinks. These ecosystems, labeled as "blue carbon" sinks, potentially sequester more than 10 times the carbon that tropical and temperate ecosystems do. Thus, conservation and restoration of these blue carbon sinks is imperative. We explored how much carbon is currently stored in dwarf red mangrove *Rhizophora* mangle biomass in tidal creeks of Eleuthera, Bahamas. In October of 2012, four sites were selected near Cape Eleuthera, maximizing site variability. All sampling was done from six plots established at each site. The quantity of carbon stored in mangroves was determined from plant biomass, which was extrapolated from plant volumes. Mangrove volumes were determined from growth parameters of individuals. It was observed that there were large differences from site to site in number of individuals, sediment depth, biomass accumulation, and carbon allocation of mangroves, but the total amount of carbon stored from site to site in mangroves did not differ. The site with the greatest biomass and carbon storage also had the greatest sediment depth, suggesting a correlation between the two. Regardless of the site to site variability, mangroves proved to be good stores for carbon. Future work should search for the factors that explain site to site variability.

Introduction

A goal for both conservation and restoration of fish habitats should be to select sites that maximize ecosystem services. In general, mangrove ecosystems provide several valuable ecosystem services. Mangroves serve as a boundary between terrestrial, estuarine, and nearshore marine ecosystems while also protecting communities from storms and coastal erosion (Lee et al. 2014). Mangroves are considered to be a foundation species supporting nurseries (Lee at al. 2014) and are heavily involved in controlling ecosystem dynamics, including fluxes of energy and nutrients, hydrology, food webs, and biodiversity. Mangrove leaf litter has also been shown to be an essential part of the diet of brachyuran crabs (Bui and Lee 2014). Serving as a home for many different species, they generate more than one billion U.S. dollars each year in ecosystem services, and it is estimated that 80% of global fish catches are in some way related to mangroves (Polidoro et al. 2010).

* Corresponding author: pdaneshg@monmouth.edu

Beyond these services, mangrove ecosystems may play a large role in mitigating the effects of climate change caused by rising levels of atmospheric CO2 by serving as a natural sinks for carbon (Alongi 2012; Donato et al. 2012; Lee et al. 2014). Mangrove ecosystems, which have been labeled as blue carbon sinks (along with sea grass beds and salt marshes), have been mostly overlooked in their role of carbon sequestration, but recent reports highlight them as highly efficient sinks (McLeod et al. 2011). Mangroves have proven to be good sinks by way of sequestering carbon within their underlying sediments and aboveground and belowground biomass that would otherwise be left as atmospheric CO_2 (McLeod et al. 2011). In the short term mangrove ecosystems store carbon in plant biomass while storing carbon in the sediment for the long term. These ecosystems trap sediment and organic carbon from internal and external riparian and oceanic sources, representing a carbon sink for a larger area (McLeod et al. 2011).

Though they are extremely important resources, mangrove forests have declined drastically. It is estimated that only half of mangrove ecosystems remain, with a loss of more than 150,000 ha a year (Alongi 2002; Spalding et al. 2010). Mangrove deforestation will result in an estimated 0.01–0.02 pentagrams C/year in global carbon emissions (Donato et al. 2011). It is thought that human activities in coastal areas may disrupt carbon sequestration by mangroves and instead cause these traditional carbon sinks to switch to a net carbon sources (McLeod et al. 2011).

The role of mangrove ecosystems of the Caribbean on the global carbon cycle is unexplored, but understanding this role and determining the amount of carbon stored in mangrove ecosystems should help warrant their conservation. This study focuses on determining how much carbon is typically stored in the biomass of dwarf red mangroves *Rhizophora mangle*, the most dominant species of mangrove in the Bahamas. In this work, multiple sites were compared to explore whether or not there were differences in carbon storage of red mangroves from site to site. With the aim of determining how much carbon is stored in mangroves, it was hypothesized that carbon storage would be variable from site to site. Should mangroves prove to be good stores of carbon, then conservation and restoration measure of these ecosystems would be further justified.

Methods

Study site

Four study sites were selected near Cape Eleuthera, Bahamas: Broad Creek (24°47.907 N, 076°17.395 W), Kemp's Creek (24°48.743 N, 076°18.444 W), Deep Creek (24°45.922 N, 076°15.725 W), and Wemyss Bight (24°43.438 N, 076°13.255 W) (Figure 1). Sampling locations included tidal creeks and flats dominated by dwarf red mangroves. Locations were chosen to maximize the variability of the environmental conditions and the characteristics within in them. These sites are characterized by sandy beaches, turtle grass *Thalassia testudium* beds, and algal plains with some sharp calcium carbonate outcroppings with the dwarf red mangroves surrounding these areas (Danylchuk et al. 2007). The sediments at all four sites consisted of a mix of calcium carbonate sand and silt mixed with mangrove-dominated decomposing litter. Maximum daily tidal range in the tidal creeks and mangroves studied on Cape Eleuthera is typically approximately 0.8 m (Murchie et al. 2013).

Two of the sites, Kemp's Creek and Broad Creek, were on the north side of the cape adjacent to Rock Sound while the other two sites, Deep Creek and Wemyss Bight, were on the south side of the cape, the Exuma Sound side of the island. Wemyss Bight appeared to be the most mature of all of the sites, with the most accumulated sediment as well as the tallest dwarf red mangrove individuals.

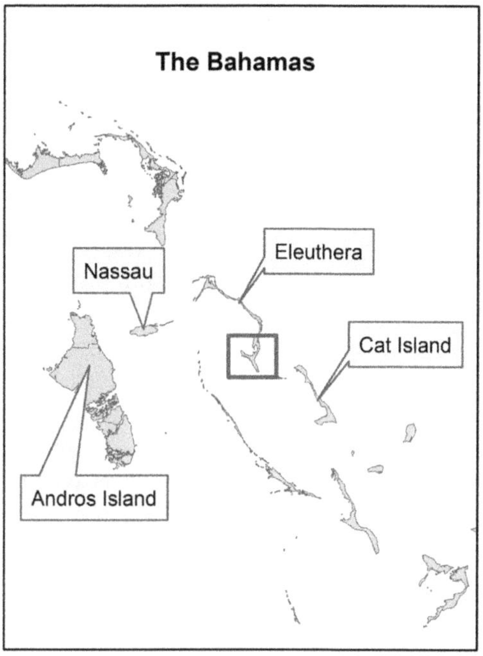

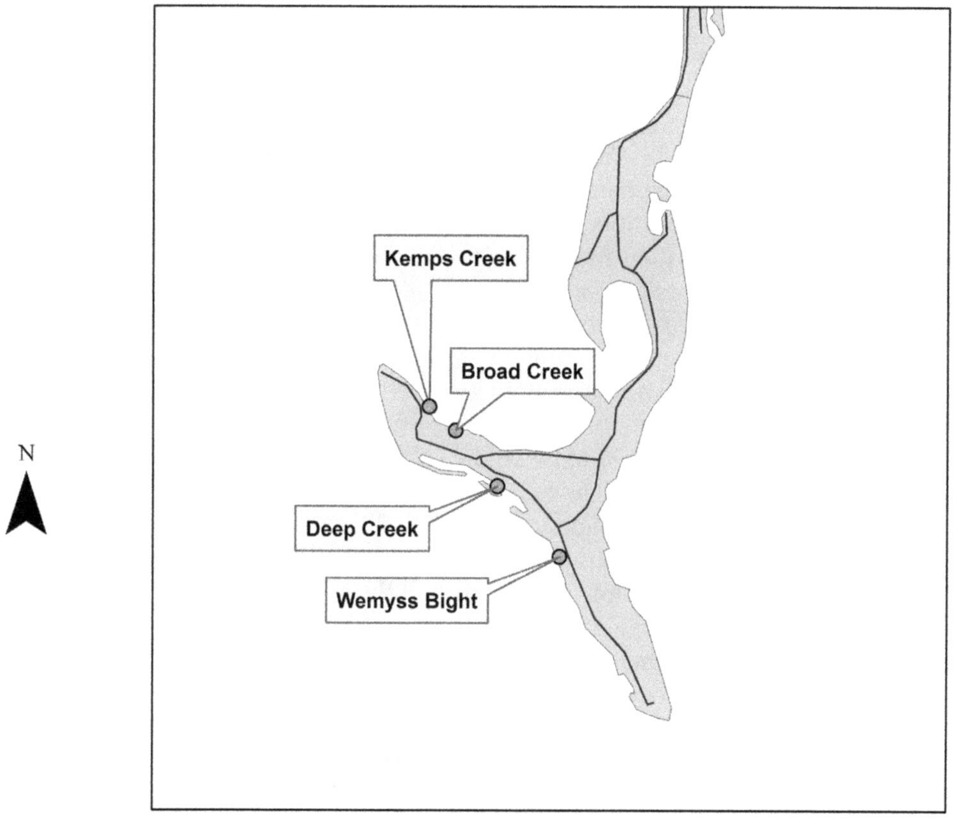

Figure 1.—Four study site locations on south Eleuthera.

A large portion of the ground at Broad Creek and Deep Creek was covered by limestone and had little sediment accumulation.

Experimental design

At each site, two 75-m transects were established parallel to each other and adjacent to the opening of the flat to the open water in a way that provided site variability. The distance between transects was different from site to site and dependent on mangrove distributions. Along each transect, three 7 × 7 m^2 plots were established at 25-m intervals (six total plots per site). Analyses were conducted comparing both sites and transects. The patchiness of mangrove flats and variable differences in densities of dwarf red mangroves throughout the flats was well represented by the transect locations.

Plant biomass sampling and carbon estimations

Data collection began in January 2013 and continued through January 2014 in three separate collection periods: January 2013, June 2013, and January 2014. Total counts of three size-classes: seedlings, samplings, and adult mangroves were quantified at each of the plots. Seedlings were identified as a single stem with no prop roots (hereafter called stilts) or branches. Saplings had one or two stilts and one or two branches. Individuals with more than two stilts were classified as adults.

Leaf counts were conducted of five individuals of each size-class of each plot. Means of these counts and the total counts of individuals per plot were used to extrapolate the total number of leaves per plot. Five fresh leaves were randomly sampled from each plot to determine mean biomass and for carbon analysis in the laboratory.

Five individuals from each class were randomly selected from each plot to estimate plant volume. Volumes of mangroves were determined by treating each mangrove as an accumulation of cylinders. Each stilt, branch, and trunk was treated as an individual cylinder for which the volume was determined using the length (l) and diameter (d) of each (volume = $\pi[d/2]^2 l$). The cylinder volumes were summed to get total volume of each mangrove individual. Volumes of individuals were extrapolated using plot counts to estimate total plot mangrove volume. Plot volumes were converted to total plot aboveground biomass using regression analyses.

Stems and stilts were sampled from each plot from each site to be used for the regression analysis. The wet volume and dry weight of each sampled stem or stilt was determined. A regression analysis was performed between stem/stilt volume and its biomass to determine whether or not volume was a good predictor of biomass. It was determined that biomass could be determined from volume using the equation $B = 3.133 + 0.032V$, where B is biomass in grams and V is volume (cm^3) at 95% confidence ($R^2 = 0.89$). The sampled stems were taken back to the laboratory for carbon analysis.

Other studies cite using diameter at breast height to estimate volume and biomass of mangrove species (Komiyama et al. 2008; Adame et al. 2013). This method was not applicable here as our study species rarely reached breast height for the measurement to be applicable and identifying a true bole proved to be challenging. Our methods of determining biomass are tedious and could be prone to errors, but our aim was to determine biomass estimates, which we believe were fairly conservative as not every branch was quantified and included in our calculations of volume and biomass.

Plant belowground biomass was determined using cores extracted from the sediment, which was extrapolated for the whole plot. Five root samples were randomly extracted at each plot using a root auger; thus, our belowground estimates of biomass are limited to the depth equivalent to the length of the auger we used. The auger used was

18 cm in length and 6.35 cm in diameter. Root samples dried for 48 h at 65°C and were then weighed. Once the root samples were weighed, the average weight for each plot was determined and used to calculate the weight per square meter based on the area of the auger and up to the depth 17.8 cm. Three of the five root samples from each plot were used in the laboratory for carbon analysis.

All leaf, stem, and root samples were then analyzed for percent carbon, percent nitrogen, and C:N ratios using a CN analyzer (Elementar CN analyzer, Germany). The analyzer works by combusting small samples (less than a gram) into small compounds that can be quantified using infrared spectroscopy. These percentages were multiplied against the calculated plot biomass numbers to estimate carbon content for the plot.

Sediment depth measurements

Sediment depth measurements were taken in June 2013 and January 2014 using a soil probe. A metal soil probe was pushed into the sediment until it was felt that the probe could not go any deeper. Usually this was when the probe reached the limestone beneath the sediment. The depth on the probe was then marked and measured. Ten random measurements were taken at each plot and averaged. The sediment measurements were averaged for the entire site in order to determine whether there was a correlation between sediment accumulation and total carbon.

Statistical analysis

Statistical analyses were performed to compare site means for sediment depth measurements, biomass and carbon values using SPSS version 18 (IBM, New York). All data were tested for normality and then normalized using log transformations when the data were nonnormal. An analysis of variance (ANOVA) was performed to compare means between sites when the data were normal (root biomass, root carbon, and sediment depth) or normalized (stem biomass, stem carbon, total biomass, and total carbon). A Tukey's post hoc analysis was used after the ANOVA to decipher which means differed from which. In cases where normality was not achieved after transformation (leaf biomass and leaf carbon), a Kruskal–Wallis test (nonparametric test) was used on the raw data with a Wilcoxon post hoc anaylsis. The statistical significance was set at $P < 0.05$ for all tests.

Results

There was a higher percentage of seedlings at each site than any other size-class (Figure 2). For all four sites, the number of individuals decreased with each size-class, from juveniles to adults. Deep Creek had the highest percentage of seedlings at 79.48%. Broad Creek had the lowest percentage of seedlings at 58.40%. Broad Creek had the highest percentage of saplings at 23.5% while Deep Creek had the lowest percentage at 14.78%. Broad Creek had the highest percentage of adults at 18.1%. Deep Creek had the lowest percentage of adults at 5.74%.

At all sites, the sediment depth changed from June 2013 to January 2014, but this change, whether an increase in depth or a decrease in depth, was never significant (Figure 3). There was a significant difference in the sediment depth among sites ($P < 0.0001$). Wemyss Bight and Kemps Creek were not statistically different in sediment depth, but they had greater mean sediment depth than Deep Creek, which was significantly greater than Broad Creek. Of all sites, Wemyss Bight had the greatest sediment depth in June 2013 (48.57 ± 8.6 cm) while the shallowest depth was observed at Broad Creek in January 2014 (6.61 ± 5.2 cm).

The greatest mean leaf biomass (g/m^2) was found at Wemyss Bight, which had significantly greater leaf biomass than Broad and Kemps creeks ($P = 0.019$, Table 1); other than Wemyss Bight, the three sites had simi-

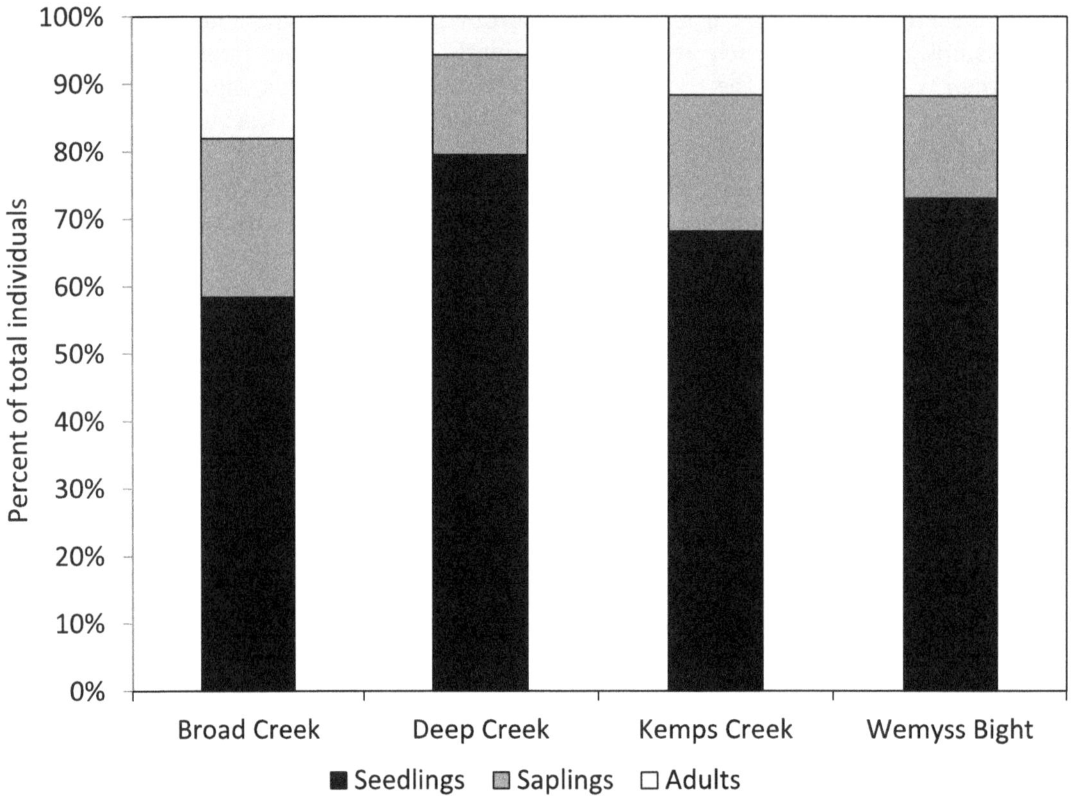

Figure 2.—Size-class distribution of dwarf red mangroves for the four study sites on Eleuthera.

lar mean leaf biomass. The greatest stem biomass (g/m^2) was found at Wemyss Bight while the lowest stem biomass was found at Kemps Creek (P = 0.029, Table 1). The greatest mean root biomass (g/m^2 up to a depth of 18 cm) was found at Wemyss Bight while the lowest mean root biomass was found at Deep Creek. No site differed statistically in mean root biomass (P = 0.17, Table 1). Wemyss Bight had the greatest total biomass (sum of leaves, stems, and roots) at 4,127.57 g/m^2. Deep Creek had the lowest total biomass at 1,819.66 g/m^2. Statistically all the sites were the same in mean total biomass (P = 0.113).

There was a strong correlation between biomass and carbon content for all biomass (r = 0.964, P < 0.001). The leaves at Wemyss Bight had the most mean carbon at 56.89 ± 46.6 g/m^2, which was significantly greater than the leaf stored carbon at Broad and Kemps Creeks (P = 0.021, Table 2). The stems at Wemyss Bight had the most carbon at a mean of 716.00 ± 553 g/m^2 while the stems at Kemps Creek had the least amount of carbon at a mean of 120.59 ± 63.6 g/m^2 (P = 0.033, Table 2). There was no significant difference in the amount of carbon stored in roots from site to site (P = 0.198, Table 2). The roots at Kemps Creek had the most carbon content with a mean of 1,099.29 ± 638 g/m^2 while the roots at Deep Creek had the least carbon content with a mean of 221.84 ± 96.6 g/m^2. There was no significant difference in the mean total carbon stored (sum of root, stems, and leaf carbon) from site to site in mangrove biomass (P = 0.125, Figure 4). Wemyss Bight had the most total carbon at 1,518.38 g/m^2. Deep Creek had the least total carbon at 655.29 g/m^2.

It was observed that there were differences from site to site in the carbon to ni-

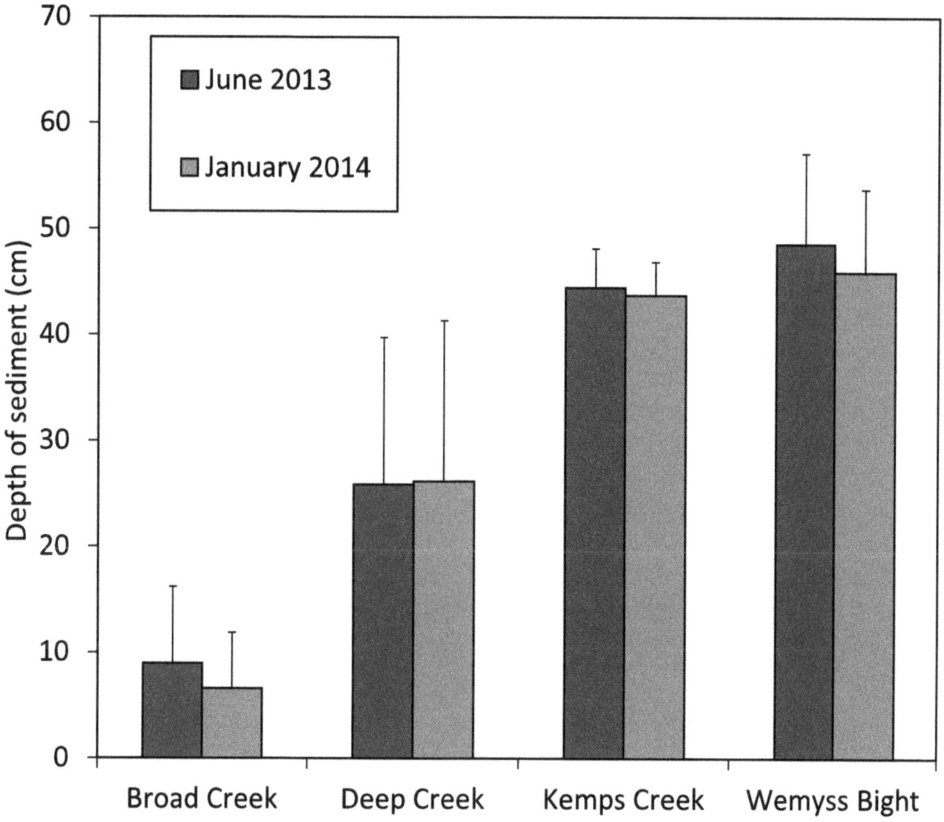

Figure 3.—Mean sediment depth measurements for the four mangrove sites on Eleuthera.

trogen ratios of the leaves and the roots (*P* = 0.004, *p* = 0.039, respectively, Table 3). Broad Creek had the highest ratios in leaves, stems, and roots with 49.03 ± 8.2, 189.86 ± 82.1 , and 58.24 ± 9.4, respectively. Kemps Creek had the lowest ratios in leaves, stems, and roots with 36.9 ± 3.9, 123.16 ± 25.4, and 46.99 ± 4.2, respectively.

Discussion

Carbon stored in mangrove biomass on Eleuthera averages 11.12 Mg/ha. Combining this with what larger quantities of carbon likely stored in the sediment verifies the value these ecosystems have as carbon sinks. This work demonstrated that there were site to site differences in class distribu-

Table 1.—Mean biomass (±SD) of leaves, stems, and roots of dwarf red mangrove (g/m^2) for the four study sites on Eleuthera. Note that significantly different means are accompanied by different letters ($P < 0.05$).

Site	Leaves	Stems	Roots
Broad Creek	27.25 ± 20.6 y	617.62 ± 434.7 yz	2,381.83 ± 2,033.8
Deep Creek	41.36 ± 13.7 yz	555.72 ± 153 yz	1,222.56 ± 1,084.6
Kemps Creek	30.64 ± 22.1 y	302.72 ± 170.4 y	3,326.18 ± 1,906.6
Wemyss Bight	126.08 ± 102.9 z	1,643.18 ± 1,258.3 z	2,358.30 ± 806.2
P-value	0.019	0.029	0.170

Table 2.—Mean carbon stored (±SD) in leaves, stems, and roots of dwarf red mangrove (g/m^2) for the four study sites on Eleuthera. Note that significantly different means are accompanied by different letters ($P < 0.05$).

Site	Leaves	Stems	Roots
Broad Creek	12.03 ± 9.2 y	258.27 ± 183.6 yz	773.37 ± 692.7
Deep Creek	19.1 ± 6.5 yz	228.26 ± 64.3 yz	221.84 ± 96.6
Kemps Creek	13.72 ± 10.0 y	120.59 ± 63.6 y	1,099.29 ± 638.0
Wemyss Bight	56.89 ± 46.6 z	716.0 ± 553.0 z	745.48 ± 247.9
P-value	0.021	0.033	0.198

tion, sediment depth, carbon allocation, and nutrient availability in dwarf red mangrove flat ecosystems on Eleuthera. How carbon is stored in the mangrove ecosystems of the Bahamas and likely the rest of the Caribbean is variable from place to place. Some of the explanations that account for this variability among sites include class distribution ratios, sediment accumulation, and potentially nutrient availability.

Across all sites, there were more seedlings than any other age-class. This result was expected, as the numbers of individuals in each class should decline as the indi-

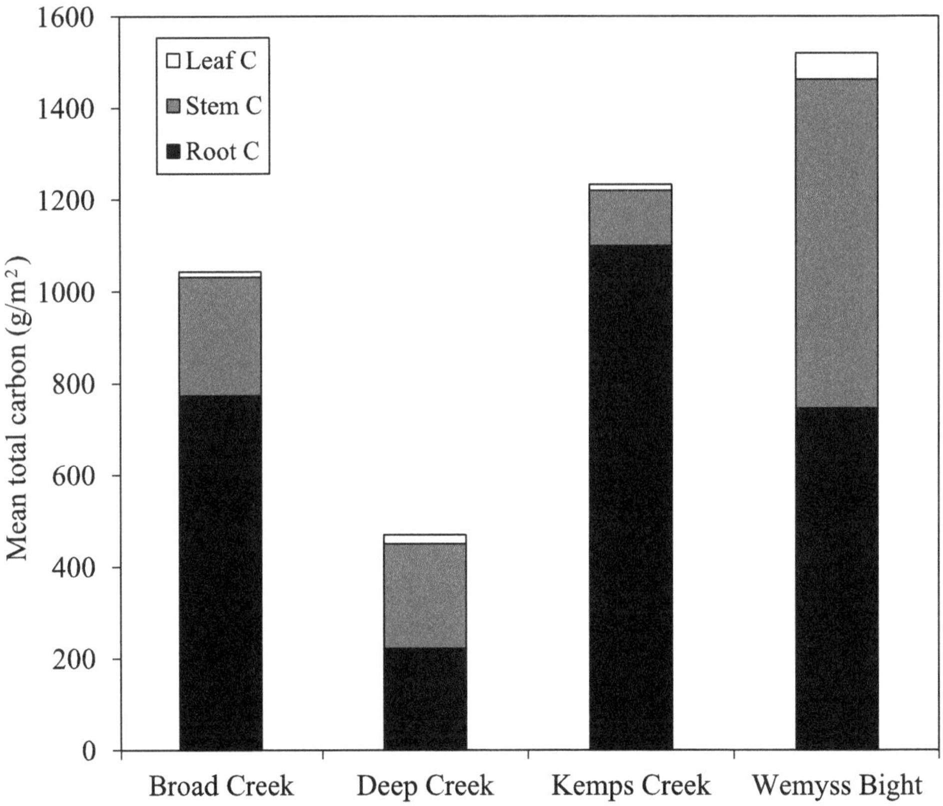

Figure 4.—Mean total carbon stored in dwarf red mangroves per square meter of the four study sites on Eleuthera. Root carbon represented up to a depth of 18 cm. There was no significant difference in mean total carbon storage from site to site ($P = 0.125$).

Table 3.—Mean C:N ratios (±SD) of leaves, stems, and roots of dwarf red mangrove for the four study sites on Eleuthera. Note that significantly different means are accompanied by different letters ($P < 0.05$).

Site	Leaves	Stems	Roots
Broad Creek	49.03 ± 8.2 z	189.86 ± 82.1	58.24 ± 9.4 yz
Deep Creek	37.45 ± 4.3 y	136.14 ± 39.4	59.57 ± 8.7 z
Kemps Creek	36.9 ± 3.9 y	123.16 ± 25.4	46.99 ± 4.2 y
Wemyss Bight	45.07 ± 5.9 yz	143.48 ± 17.2	57.81 ± 8.3 yz
P-value	0.004	0.304	0.039

viduals mature. As mangroves mature, the probability of death increases with age and size due to intraspecific competition and self-thinning, which is why there are fewer adults than seedlings or saplings at each site (Rajkaran and Adams 2012). Resource demands also increase as the mangrove individuals mature; thus, limiting resources then become a thinning factor. Light is not a limiting factor for establishing seedlings, even in dense mangrove, as dwarf red mangrove has been shown to have traits that allow it to be both light demanding and shade tolerant (Farnsworth and Ellison 1996).

Our estimates of carbon storage in seedlings thus are values for a specific moment in time and may not project into future carbon estimates. It was observed that Deep Creek had the most number of seedlings, which suggests great potential for high productivity and carbon storage in the future as those individuals mature. Broad Creek had the most adults and the lowest number of seedlings. This may imply that either the site is mature or that it is a poor site for seedling recruitment. This site had a lot of bedrock and very little sediment accumulation, which is necessary for the establishment of new individuals.

The depth of the sediment changed from June 2013 to January 2014 at all sites. Deep Creek increased in sediment depth while the other three sites decreased in sediment depth, but none of these changes were statistically significant. These changes in sediment depth were expected as the tides move sediment in and out of the creeks as has been observed globally at several sites in several coastal ecosystems (Gourgue et al. 2013). These changes in depth would most likely become significant after longer periods of time. We expect that all sites would accumulate large quantities of sediment after several years. Mangroves have proven to be quite valuable in regard to sediment accumulation contributing to inorganic sedimentation, soil development, vertical land building, and stabilization (Lee et al. 2014).

The depth of the sediment may have been tied partially to mangrove productivity and carbon accumulation. High productivity has been reported to increase sediment accumulation rapidly (Osland et al. 2012). Wemyss Bight had the largest individual mangroves and the greatest sediment depth, suggesting a correlation between mangrove biomass and sediment depth: the larger the individual, the more sediment they will accumulate (Osland et al. 2012). An increase in sediment depth should bring in an influx of nutrients, which could pulse productivity and biomass accumulation. If large mangroves trap more sediment and more sediment brings more nutrients, which promote mangrove biomass production (Duarte et al. 1998), then there is a two-way positive feedback mechanism, where sediment accumulation and mangrove productivity fuel each other. This feedback cycle does not seem apparent for Broad Creek, which had the lowest sediment depth but mean total biomass comparable to the other sites. The sediment depth values at Broad Creek were due to abundant areas of bare limestone rock, which were recorded as a depth of zero whenever the limestone rock areas were randomly sampled for sediment depth measurements. Disregarding these bare areas, the sediment depth mean for the site would be greater and potentially linked to mangrove growth.

The least amount of biomass was stored in the leaves and the average leaf weight differed across sites. Leaf texture and average size are variable in Rhizophoraceae, which could potentially account for the differences in average weight (Tomlinson et al. 1979). High leaf production is important as it would feed back to greater overall mangrove productivity by being the location of photosynthesis (Medeiros and Sampaio 2011). Nutrient inputs most likely influenced leaf biomass. This would have to be explored in a study examining nutrient availability of these sites in the sediment.

The least amount of total carbon was stored in the leaves as compared to the stems and roots. However, leaves had the highest percentage of carbon values across all sites compared to the roots, which had the lowest percentage of carbon values across all sites. Thus, there is value in leaf production in terms of serving as good carbon sinks. At the same time, roots have a low percentage of carbon, but because of the high volume/biomass of overall roots, roots remain the largest pool of carbon of mangrove biomass.

Most of the mangrove biomass was in the formation of roots, which was also observed in the Florida Everglades (57–78% of the total productivity was in root production; Castaneda-Moya et al. 2011). Across all sites, roots were only sampled to a depth of 18 cm using an auger. Some of the sites had three times that depth in sediment, meaning that not all of the root biomass is reported here, and there may be significantly more root biomass. How much root biomass accumulated in the nonmeasured sediment depths would be dependent on root penetration, which can be dictated by several factors, including waterlogging, hypoxia, and sediment compactness (McKee 1996). Thus, it is unlikely that roots were consistent through the sediment depth. Root production was most likely tied to high nutrient availability that comes through sediment accumulation (Castaneda-Moya et al. 2013). The C:N data presented here suggests there may be potentially differences in nutrient availability, which would have to be determined from further sediment analysis. High C:N could result from high nitrogen in the sediment or low carbon capture rates via photosynthetic rates.

There are some possible explanations as to why the greatest productivity and carbon accumulation were observed at Wemyss Bight. First, Wemyss could simply be the oldest site and has had the most time to accumulate biomass. Aging mangroves in tropical areas can be difficult as they do not form annual rings like most deciduous woody species found in temperate environments. Beginning with this work, we have been monitoring growth of the marked mangrove individuals over time. We should be able to use these to calculate growth rates, which can be used to determine how old the individuals are assuming constant growth rates. Afterwards, site differences may be explained by age.

Freshwater input could also potentially explain why Wemyss Bight had the most productivity. Several works have tied mangrove productivity to freshwater availability. As salinity increases, the stature of mangroves decline (Castaneda-Moya et al. 2013). In the Middle East, a study found that groundwater was the major contributor to sustaining mangrove density (Drexler and Carlo 2002). The sites in Eleuthera are highly saline environments; further studies would need to be done to determine if there are freshwater inputs at the more productive sites.

Our results indicate that mangroves in the Caribbean are good carbon sinks and have the potential to help mitigate the effects of global climate change by reducing atmospheric CO_2. This picture will be clearer when belowground carbon storage in the sediment has been described in future work. Comparing to other mangrove ecosystems, the dwarf red mangrove flats ecosystems of the Bahamas may not hold as much carbon per hectare as observed in other mangrove ecosystems (Alongi 2012; McLeod et al. 2011); however, they are just as valuable when compared to terrestrial ecosystems and when all of the ecosystem services they provide (Lee et al. 2014) are considered.

There is some evidence that mangrove ecosystems, particularly those along islands, function as sinks and sources (exporters) of carbon through consumption of mangrove leaf litter by migrating detritivores and washouts caused by large rainfall events (Bui and Lee 2014; Lee et al. 2014). Rainfall events that could increase in frequency with climate change in tropical areas could destroy the value of mangroves as carbon sinks. Sea level rise associated with climate change also threaten mangrove ecosystems as islands are squeezed, leading to losses in suitable coastline for mangrove establishment (Donato et al. 2012). Tragically, the mitigating effects of the mangroves as carbon sinks may not prevent the climate change events that will alter their ecosystem services and deprive mangroves of suitable areas to establish on tropical islands.

Acknowledgments

We would like to thank the Fisheries Conservation Federation and Monmouth University Urban Coast Institute for their financial support. Thanks to Carleigh Engstrom and Mitchell Mickley for their support in the laboratory. Thanks to Richard Bastian, Lauren Johnson, Kaitlyn Kramer, and Alex Moncman for guidance on statistical analyses.

References

Adame M. F., J. B. Kauffman, I. Medina, J. N. Gamboa, O. Torres, J. P. Caamal, M. Reza, and J. A. Herrera-Silveira. 2013. Carbon stocks of tropical coastal wetlands within the karstic landscape of the Mexican Caribbean. PLOS One 8(2):e56569.

Alongi, D. M. 2002. Present state and future of the world's mangrove forests. Environmental Conservation 29:331–349.

Alongi, D. M. 2012. Carbon sequestration in mangrove forests. Carbon Management 3:313–322.

Bui, T. H. H., and S. Y. Lee. 2014. Does "you are what you eat' apply to mangrove grapsid crabs? PLOS ONE 9(2):e89074.

Castaneda-Moya, E., R. R. Twilley, and V. H. Rivera-Monroy. 2013. Allocation of biomass and net primary productivity of mangrove forests along environmental gradients in the Florida Coastal Everglades, USA. Ecosystems 307:226–241.

Castaneda-Moya, E., R. R. Twilley, V.H. Rivera-Monroy, B.D. Marx, and C. Coronado-Molina. 2011. Patterns of root dynamics in mangrove forests along environmental gradients in the Florida Coastal Everglades, USA. Ecosystems 14:1178–1195.

Danylchuk, A. J., S. E. Danylchuk, S. J. Cooke, T. L Goldberg, J. B. Koppelman, and D. P Phillip. 2007. Post-release mortality of Bonefish, Albula vulpes, exposed to different handling practices during catch-and-release angling in Eleuthera, The Bahamas. Fisheries Management and Ecology 14:149–154.

Donato, D. C., J. B. Kauffman, R. A. Mackenzie, A. Ainsworth, and A. Z. Pfleeger. 2012. Whole-island carbon stocks in the tropical Pacific: implications for mangrove conservation and upland restoration. Journal of Environmental Management 97:89–96.

Donato, D. C., J. B. Kauffman, D. Murdiuarso, S. Kurniato, M. Stidham, and M. Kanninen. 2011. Mangroves among the most carbon-rich forests in the tropics. National Geosciences 4:293–97.

Drexler, J. Z., and E. W. DeCarlo. 2002. Source water partitioning as a means of characterizing hydrologic function in mangroves. Wetlands Ecology and Management 10:103–113.

Duarte, C. M., O. Geertz-Hansen, U. Thampanya, J. Terrados, M. D. Fortes, L. Kamp-Nielsen, J. Borum, and S. Boromthanarath. 1998. Relationship between sediment conditions and mangrove Rhizophora apiculata seedling growth and nutrient status. Marine Ecology Congress Series 175:277–283.

Farnsworth, E. J., and A. M. Ellison. 1996. Sunshade adaptability of the red mangrove, *Rhizophora mangle* (Rhizoporaceae): changes through ontongeny at several levels of biological organization. American Journal of Botany 83:1131–1143.

Gourgue, O., W. Baeyens, M. S. Chen, A. DeBrauwere, B. DeBrye, E. Deleersnijder, M. Elskens, and V. Legat. 2013. A depth-averaged two-dimensional sediment transport model for environmental studies in the Scheldt estuary and tidal river network. Journal of Marine Systems 128:27–39.

Komiyama, A., J. E. Ong, and S. Poungparn. 2008. Allometry, biomass, and productivity of mangrove forests: a review. Aquatic Botany 89:128–137.

Lee, S. Y., J. H. Primavera, F. Dahdouh-Geubas, K. McKee, J. O. Bosire, S. Cannici, K. Diele, N. Koedam, C. Marchand, I. Mendelssohn, N. Mukherjee, and S. Record. 2014. The ecological role and ecosystem services of tropical mangroves: a reassessment. Global Ecology and Biogeography 23:726–743.

McKee, K. L. 1996. Growth and physiological responses of neotropical mangrove seedlings to root zone hypoxia. Tree Physiology 16:883–889.

McLeod, E., G. L. Chmura, S. Bouillon, R. Salm, M. Bjork, C. M. Duarte, C. E. Lovelock, W. H. Schlesinger, and B. R. Silliman. 2011. A blueprint for blue carbon: toward an improved understanding of the role of vegetated coastal habitats in sequestering CO_2. Frontiers in Ecology and Environment 9:552–560.

Medeiros T. C., and E. Sampaio. 2013. Leaf and flower formation in shoot tips of mangrove trees in Pernambuco, Brazil. Wetlands Ecology Management 21:209–217.

Murchie, K. J., S. J. Cooke, A. J. Danylchuk, S. E. Danylchuk, T. L. Goldberg, C. D. Suski, and D. P. Phillip. 2013. Movement patterns of Bonefish (*Albula vulpes*) in tidal creeks and coastal waters of Eleuthera, The Bahamas Fisheries Research 147:404–412.

Osland M. J., A. M. Spivak, J. A. Nestlerode, J. M. Lessmann, A. E. Almario, P. T. Heitmuller, M. J. Russell, K. W. Krauss, F. Alvarez, D. D. Dantin, J. E. Harvey, A. S. From, N. Cormier, and C. L. Stagg. 2012. Ecosystem development after mangrove wetland creation plant-soil change across a 20-year chronosequence. Ecosystems 15:848–866.

Polidoro B. A., K. E. Carpenter, L. Collins, N. C. Duke, A. M. Ellison, J. C. Ellison, E. J. Farnsworth, E. S. Fernando, K. Kathiresan, N. E. Koedam, S. R. Livingstone, T. Miyagi, G. E. Moore, V. N. Nam, J. E. Ong, J. H. Primavera, S. G. Salmo, III, J. C. Sanciangco, S. Sukardjo, Y. Wang, and J. W. Hong Yong. 2010. The loss of species: mangrove extinction risk and geographic areas of global concern. PLOS ONE 5(4):e10095.

Rajkaran, A., and J. Adams. 2012. The effects of environmental variables on mortality and growth of mangroves at Mngazana Estuary, Eastern Cape, South Africa. Wetlands Ecology Management 20:297–312.

Spalding M. D., M. Kainuma, and L. Collins. 2010. World atlas of mangroves. Earthscan, London.

Tomlinson, P. B., R. B. Primack, and J. S. Bunt. 1979. Preliminary observations on floral biology in mangrove Rhizophoraceae. Biotropica 11:256–277.

American Fisheries Society Symposium 83:113–116, 2015

EXTENDED ABSTRACT

Effects of an Episodic Drought on a Floodplain Subsidy Consumed by Mangrove River Fishes

Ross E. Boucek* and Jennifer S. Rehage
Southeast Environmental Research Center
Department of Earth and Environment, Florida International University
11200 SW 8th Street, Miami, Florida 33199, USA

Introduction

Ecosystem connectivity is vital to provisioning mangrove-dependent fishes (Mumby 2006). In mangrove rivers, piscivorous fish populations are often subsidized by seasonal, highly abundant prey that flux into rivers from adjacent floodplain habitats at the onset of the dry season. The magnitude of these floodplain prey subsidies may be dependent on (1) rainfall dynamics and inundation patterns of floodplains during the previous flood season (i.e., the growing season for the forage species), and (2) the population dynamics of floodplain predators that may regulate forage fish population dynamics within the floodplain habitat (Welcomme and Halls 2004). Extreme and episodic droughts can disrupt these seasonal prey subsidies by reducing the duration of the flood season, resulting in changes to the population dynamics of floodplain forage fish, as well as affecting floodplain predators. Thus, in dry seasons following these disturbances, the quantity and composition of these subsidies may change as floodplain communities recover from disturbance. However, the long-term effects of droughts on tropical river fish floodplain subsidies remains understudied.

In the coastal mangrove rivers of the subtropical Everglades (U.S.), seasonal rainfall patterns dictate floodplain prey subsidies to a coastal mangrove river piscivore, Common Snook *Centropomus undecimalis* (Boucek and Rehage 2013). In 2011, a once-in-10-years drought affected South Florida. The drought caused floodplains to remain dry for 100 d, tripling the average number of dry days per year. This drought had profound impacts on floodplain fishes and likely altered subsidies to Common Snook (Boucek and Rehage 2014). The objective of this study was to determine the impact and legacy of the 2011 drought on floodplain subsidies. We hypothesized that 1 year after the drought, due to the reduced duration of the flood season, the floodplain subsidies would be lessened. Two years after the drought, due to the slower recovery of floodplain piscivores and top-down release of forage fishes on the floodplain, we expected a temporary increase in the magnitude of the subsidy to Common Snook.

Methods

We conducted this study in the oligohaline reaches of the Shark River in Everglades National Park. We sampled five oligohaline creeks with three replicate transects at each creek monthly between December and June, 2010 to 2013. We sampled both piscivores and prey using a boat-mounted electrofisher (285 samples total). We report electrofishing catch per unit effort (CPUE) as the number of fish caught per 100 m of shoreline ([CPUE/

* Corresponding author: rbouc003@fiu.edu

distance traveled]/100 m; Boucek and Rehage 2013).

Floodplain piscivores

We tracked abundance changes in three piscivore species, Florida Largemouth Bass *Micropterus salmoides floridanus*, currently listed as a subspecies of Largemouth Bass *M. salmoides*, Bowfin *Amia calva*, and Common Snook. Florida Largemouth Bass and Bowfin are floodplain piscivores and could regulate floodplain prey populations, while Common Snook are a river resident (see Boucek and Rehage 2013). To quantify the effects of drought on floodplain piscivore abundance and average body size, we compared effects across the 3 years using repeated measures analysis of variances, with transect nested within month as subjects (abundance) and one-way analysis of variance, respectively (body size). Tukey's pairwise comparisons were used in post hoc tests and conducted tests in SAS 9.2 (SAS Institute, Rockville, Maryland).

Floodplain prey

Floodplain forage fishes consist of four centrarchid species (Boucek and Rehage 2013). These forage fishes pulse into mangrove rivers in the winter and spring, coinciding with floodplain drying (Boucek and Rehage 2013). Two cohorts of centrarchids enter the river system, early-spawned fishes ranging from 8 to 15 cm standard length (SL) and later-spawned centrarchids less than 8 cm SL. We calculated the average biomass per 100 m of floodplain forage fishes that entered the river. These values combined with monthly electrofishing samples were used to develop biomass curves for each year of the study. With these curves, the relative magnitude of the subsidy to Common Snook was determined by calculating the area under these curves. We partitioned these biomass curves by the larger and the smaller cohorts. We then compared the area under these curves across years using chi-square tests.

Results

Floodplain piscivores

Both floodplain piscivores' abundances and average body size varied across years (Florida Largemouth Bass: $F > 8.9$, $p < 0.01$; Bowfin $F > 12.6$, $p < 0.01$), whereas Common Snook's abundance and size structure did not differ annually ($F < 1.1$; $p > 0.21$). Both Florida Largemouth Bass and Bowfin abundance decreased by 61% and 81% 1 year after the drought ($p < 0.03$; Figure 1A). However, contrary to our predictions, 2 years following the drought, Florida Largemouth Bass abundance returned to predisturbance values ($p = 0.15$), while Bowfin abundance showed a 2.3-fold increase from the predisturbance year ($p < 0.01$; Figure 1A). Average Florida Largemouth Bass length increased by 8% ($p = 0.02$) in 2012 while average Bowfin length did not change ($p = 0.32$). In 2013, Florida Largemouth Bass and Bowfin average body length decreased by 12% and 9% ($p < 0.05$; Figure 1B).

Floodplain prey

During the predisturbance year, larger-bodied sunfishes made up a larger portion of the subsidy than did smaller sunfishes, accounting for 62% of the total biomass ($\chi^2 = 5.2$, $p = 0.02$; Figure 2). One year after the drought, total floodplain biomass entering the river decreased by 58% ($\chi^2 = 16.6$, $p < 0.01$; Figure 2). Despite decrease in biomass fluxing into the system, the ratio of large sunfishes to small sunfishes was similar to predisturbance (36% small versus 64% large sunfishes, $\chi^2 = 7.6$, $p = 0.01$). Two years after the drought, floodplain prey biomass increased 3.6-fold compared to the predisturbance year. The identity of this subsidy also changed, with now 97% of the subsidy composed of early-spawned larger sunfishes (Figure 2).

Conclusions

Our results suggest that drought lessened floodplain prey biomass entering a man-

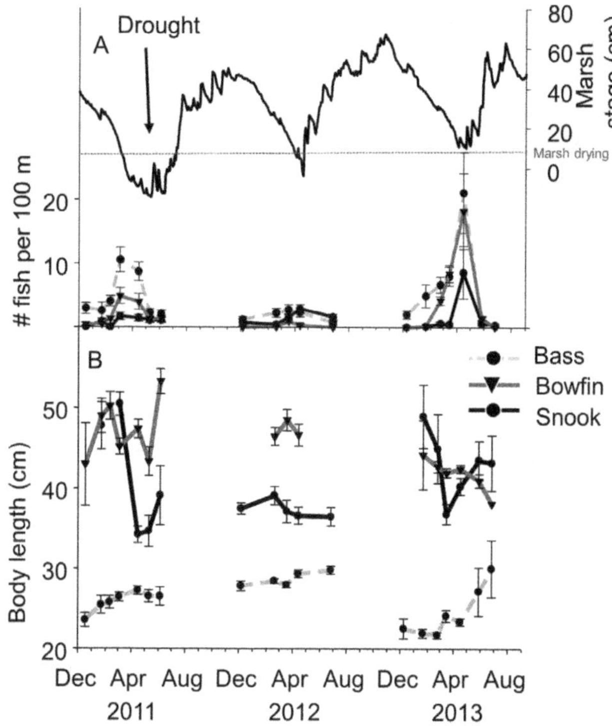

Figure 1.—(A) Variation in electrofishing catch per unit effort (number of fish per 100 m of mangrove shoreline) of piscivores Florida Largemouth Bass (gray dashed line), Bowfin (gray solid line), and the river piscivore Common Snook (black solid line) across the 3 years of study. The black line in panel A shows daily marsh floodplain stage. The dashed horizontal line represents when marsh stage reaches levels that are unavailable to floodplain fishes. Panel B shows variation in body length of piscivores.

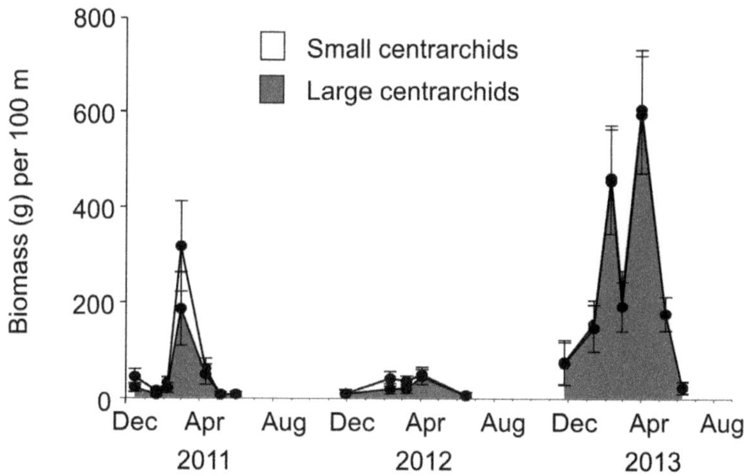

Figure 2.—Variation in floodplain biomass entering the coastal river partitioned by early-spawned centrarchids (gray area under the curve) and late-spawned centrarchids (white area under the curve across the years of study.

grove river 1 year after the disturbance. This result corroborates findings in Boucek and Rehage (2014) that found 60% decreases in sunfish abundance across a wide spatial distribution of estuarine creeks following the drought. We did find some support for our predation release hypothesis 2 years after the drought. Even though predator abundance recovered, average floodplain predator body size was skewed toward smaller individuals that may be less efficient at consuming larger-bodied earlier-spawned sunfishes. Thus, due to the potential top-down release on the early-spawned cohort, prey biomass entering the river tripled and largely consisted of earlier-spawned larger sunfishes. If droughts become more frequent under future climate change scenarios, we may expect that the year-to-year predictability of floodplain subsidies to decrease, which will likely reduce the stability of mangrove fisheries that rely on these resources.

References

Boucek, R. E., and J. S. Rehage. 2013. No free lunch: displaced marsh consumers regulate a prey subsidy to an estuarine consumer. Oikos 122:1453–1464.

Boucek, R. E., and J. S. Rehage. 2014. Climate extremes drive changes in functional community structure. Global Change Biology 20:1821–1831.

Mumby, P. J. 2006. Connectivity of reef fish between mangroves and coral reefs: algorithms for the design of marine reserves at seascape scales. Biological Conservation 128:215–222.

Welcomme, R. L., and A. S. Halls. 2004. Dependence of tropical river fisheries on flow. Pages 267–283 in R. L. Welcomme and T. Petr, editors. Proceedings of the second international symposium on the management of large rivers for fisheries, volume 2. Food and Agriculture Organization of the United Nations & Mekong River Commission, FAO Regional Office for Asia and the Pacific, RAP Publication 2004/17, Bangkok.

American Fisheries Society Symposium 83:117–121, 2015

EXTENDED ABSTRACT

Mangrove Habitat Use by Resident versus Migrant Sport Fishes in the Coastal Everglades

Jennifer S. Rehage*, Ross E. Boucek, and Jessica A. Lee
Southeast Environmental Research Center
Department of Earth and Environment, Florida International University
11200 SW 8th Street, Miami, Florida 33199, USA

Introduction

Mangroves are well known for providing valuable nursery habitat for a myriad of estuarine and marine fishes (Faunce and Serafy 2006; Murchie and Daneshgar 2015). Yet, their role as refuge habitats across the freshwater–estuarine interface is understudied. We refer to refuge habitats as those that provide resistance (survival) and/or resilience (recovery time) to populations from the effects of disturbance, such as drying events (Parkos et al. 2011).

In the Florida Everglades, the recurrent pattern of seasonal drying of freshwater marshes strongly structures fish communities, limiting abundance and affecting distribution across the landscape (Parkos et al. 2011; Boucek and Rehage 2013). This marked seasonality is magnified by drainage and impoundment over the past 100 years (McVoy et al. 2011). Freshwater inflows into the southern Everglades have been greatly reduced, resulting in a higher frequency of drying in upstream marshes and higher salinity conditions in downstream mangroves. In response to these drier conditions, freshwater fishes move from marshes to deeper refuge habitats, including mangrove areas (Boucek and Rehage 2013); however, how these migrant fishes use mangroves relative to resident estuarine taxa remains poorly understood.

We compared the use of mangrove habitats by migrant Florida Largemouth Bass *Micropterus salmoides floridanus* to that of resident Common Snook *Centropomus undecimalis*. Both species are important sport fishes, but Florida Largemouth Bass are primarily freshwater migrants and highly abundant in mangroves only in the dry season, whereas Common Snook reside in mangroves year-around (leaving to spawn at sea, Boucek and Rehage 2013). We compared their movements and seasonal patterns of distribution in meso to oligohaline mangrove reaches. We expected more pronounced seasonality in distribution in the migrant species because of their immigration in and out of the system in relation to marsh drying. Since Florida Largemouth Bass are intolerant of high salinities and not highly mobile, we hypothesized a more limited distribution and lower movement relative to the euryhaline Common Snook.

Methods

We sampled Florida Largemouth Bass and Common Snook in the upper 14 km of the Shark River estuary in the southwestern Everglades. The Shark River estuary is a coastal river that connects Shark River Slough, the main freshwater drainage of the southern

* Corresponding author: rehagej@fiu.edu

Everglades, to extensive riverine mangroves downstream (McVoy et al. 2011). The focal species were sampled repeatedly at 14 fixed sites located along an upstream to downstream gradient (salinity = 2.4 ± 0.2 parts per thousand, depth = 1.2 ± 0.02 m, Figure 1A). To document seasonal distributions, we sampled in the wet (November–December), early dry (February–March), and late dry seasons (April–May) in 2010–2013.

Standardized sampling was conducted via boat electrofishing along shorelines, which were dominated by red mangroves throughout (n = 348 samples). We report

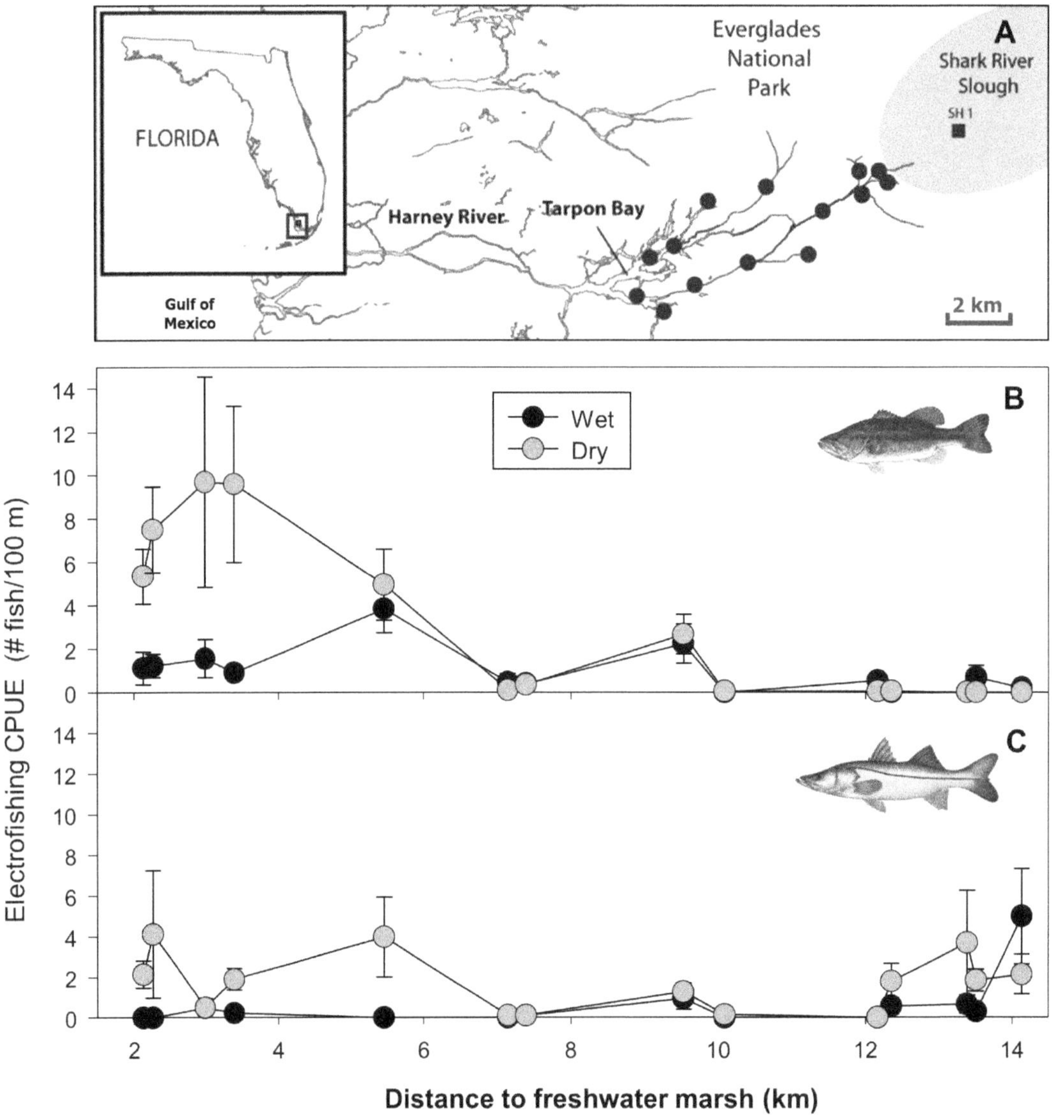

Figure 1.—(A) Map of the 14 mangrove sites in the upper Shark River estuary, and spatial variation in catch per unit effort for (B) migrant Florida Largemouth Bass and (C) resident Common Snook across seasons and as a function of distance to freshwater marshes upstream. We used distance to hydrostation SH1 located in a freshwater marsh (in A) as a proxy for distance to marsh habitats.

catch per unit effort (CPUE) as the number of Florida Largemouth Bass and Common Snook caught per 100 m of shoreline ([CPUE/distance traveled]*100 m; see Boucek and Rehage 2013 for details). Upon capture, fish were counted and those greater than 120 mm were implanted with passive integrated transponder (PIT) tags in the abdominal cavity as unique identifiers and then released. PIT-tag deployment continued through March 2014 with additional seasonal electrofishing, totaling 1,800 PIT tags deployed on Florida Largemouth Bass and 769 on Common Snook. All captured fish were scanned to check for PIT tags, and additional recaptures were obtained from anglers equipped with readers. We used capture–recapture Euclidean distances to determine fish movement patterns (e.g., Paukert et al. 2006). We examined variation in CPUE (log transformed) and recapture distances with generalized linear models in SAS 9.2 (SAS Institute, Rockville, Maryland). We tested for variation in recapture distances between species, and for CPUE, we tested for the effects of species, seasons (wet versus the average of early and late dry), distance (0–14 km away from marshes), and interactions. Preliminary analysis showed little difference between the two dry-season samples.

Results and Discussion

Despite being a migrant species, Florida Largemouth Bass were 80% more abundant in mangrove areas than resident Common Snook (2.4 versus 1.3 fish/100 m; $F_{1,642}$ = 8.5, p = 0.004). Both species were also more abundant in the dry season, with 84% of the Florida Largemouth Bass and 87% of the Common Snook caught during the dry season ($F_{1,642}$ = 23.7, p = 0.0001, Figure 1). These dry-season increases in CPUE, however, were not equally distributed across space (three-way interaction: $F_{13,642}$ = 2.1, p = 0.014). Florida Largemouth Bass increased sevenfold and Common Snook threefold between the wet and dry samples in the upper reaches of the estuary (4–5 km from marshes). Downstream, Florida Largemouth Bass were in consistently lower abundances across both seasons, and virtually absent below 9 km. In contrast, Common Snook also increased at the downstream end of the gradient (12–14 km) in the dry season (fourfold relative to the wet season).

A total of 114 Florida Largemouth Bass and 28 Common Snook recaptures (6.3% and 3.6% recapture rate, respectively) were recorded (Figure 2). Of these, 77 Florida Largemouth Bass and 19 Common Snook were recaptured via electrofishing while 37 Florida Largemouth Bass and 9 Common Snook were recaptured by anglers. Mean distance between mark and recapture was three times greater for Common Snook than Florida Largemouth Bass, 3.3 versus 0.9 km ($F_{1,140}$ = 28.1, p = 0.0001). For Florida Largemouth Bass, 68% of recaptures were recorded within 0.5 km of where captured, whereas only 21% of Common Snook recaptures were collected at this small distance. The largest distance between mark and recapture was 12 km for Common Snook and only 7 km for Florida Largemouth Bass.

Overall, use of the mangrove region by migrant Florida Largemouth Bass and resident Common Snook varied across the upper Shark River estuary, and temporally between seasons. Common Snook were more abundant in the dry season, and these increases in abundance occurred both at the upper and lower ends of the 14-km transect. For Florida Largemouth Bass, use of the mangrove region appeared limited to the upstream reaches, as evident by the fact that Florida Largemouth Bass were collected only in the upper 9 km, and in the dry season, became highly abundant only at the headwaters (<4 km). This limited use of this mangrove refuge habitat by Florida Largemouth Bass could be due to lower habitat suitability downstream and/or dispersal limitation, and further work is needed to distinguish between the two. Salinities reached the 12

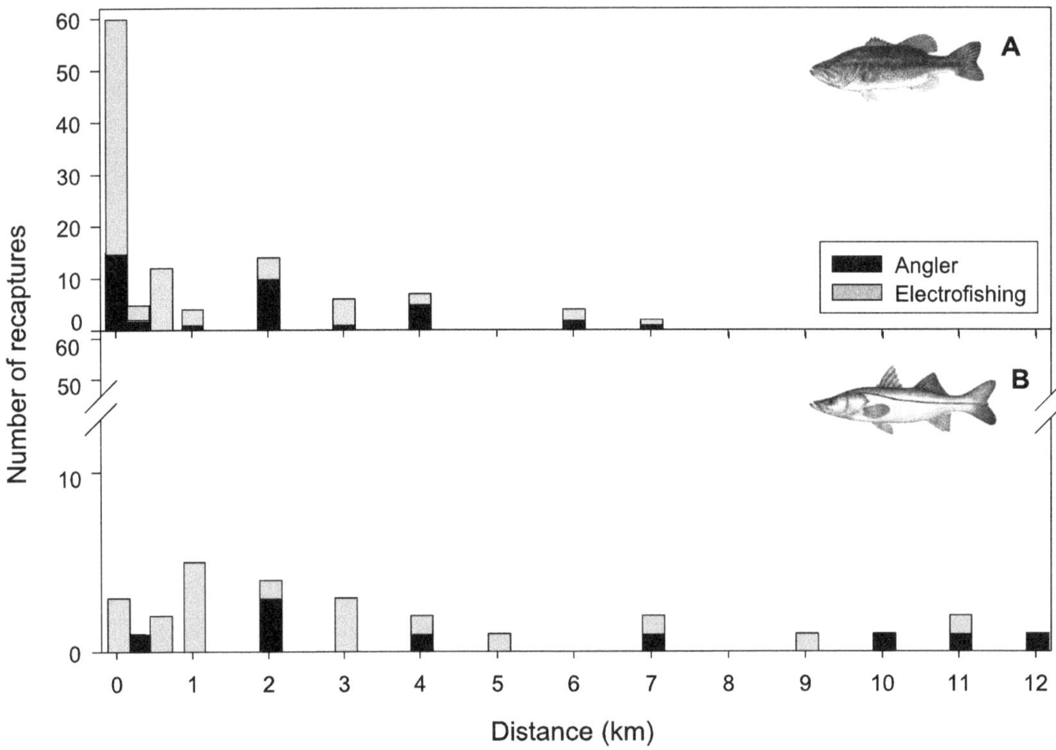

Figure 2.—Distance between capture and recapture for (A) migrant Florida Largemouth Bass (n = 114 recaptures) and (B) resident Common Snook (n = 28 recaptures) in the upper Shark River estuary. Recaptures are shown separately for electrofishing sampling (60% of recaptures) versus those provided by recreational anglers (40% of recaptures).

parts per thousand lethal limit of Florida Largemouth Bass (Norris et al. 2010) at the peak of the dry season at downstream sites. Mark–recapture results suggest lower movement for Florida Largemouth Bass relative to Common Snook, which agrees with previous Everglades work showing that a majority of radio-tagged Florida Largemouth Bass move less than 1 km (Parkos and Trexler 2014). These results highlight the importance of low-salinity mangrove regions for recreational fishes, both resident and migrant taxa. We expect habitat suitability of the mangrove zone, and thus the importance of this habitat to increase for Florida Largemouth Bass with ongoing Everglades restoration, which will decrease the frequency of marsh drying and increase oligohaline conditions in mangroves.

References

Boucek, R. E., and J. S. Rehage. 2013. No free lunch: consumers from a donor marsh community regulate a prey subsidy to an estuarine consumer. Oikos 122:1453–1464.

Faunce, C. H., and J. E. Serafy. 2006. Mangroves as fish habitat: 50 years of field studies. Marine Ecology Progress Series 318:1–18.

McVoy, C. W., W. P. Said, J. Obeysekera, J. Van Arman, and T. W. Dreschel. 2011. Landscapes and hydrology of the predrainage Everglades. University Press of Florida, Gainesville.

Murchie, K. J., and P. P. Daneshgar. 2015. Mangroves as fish habitat. American Fisheries Society, Symposium 83, Bethesda, Maryland.

Norris, A. J., D. R. DeVries, and R. A. Wright. 2010. Coastal estuaries as habitat for a freshwater fish species: exploring population-level effects of salinity on Largemouth Bass. Transactions of the American Fisheries Society 139:610–625.

Parkos, J. J., C. R. Ruetz, and J. C. Trexler. 2011. Disturbance regime and limits on benefits of refuge use for fishes in a fluctuating hydroscape. Oikos 120:1519–1530.

Parkos, J. J., and J. C. Trexler. 2014. Origins of functional connectivity in a human-modified wetland landscape. Canadian Journal of Fisheries and Aquatic Sciences 71:1418–1429.

Paukert, C. P., L. G. Coggings, and C. E. Flaccus. 2006. Distribution and movement of Humpback Chub in the Colorado River, Grand Canyon, based on recaptures. Transactions of the American Fisheries Society 135:539–544.

American Fisheries Society Symposium 83:123–125, 2015

EXTENDED ABSTRACT

Processes Affecting Movement and Survival of a Juvenile Fish in Mangrove Creeks

Andrew B. Barbour*
Program for Fisheries and Aquatic Sciences
School of Forest Resources and Conservation, University of Florida
Post Office Box 110600, Gainesville, Florida 32611, USA

Aaron J. Adams
Florida Institute of Technology, Biological Sciences
Vero Beach Marine Laboratory
150 West University Boulevard, Melbourne, Florida 32901, USA

Despite an increasing recognition of the ecological and economic values of estuarine and coastal ecosystems, we still lack a complete understanding of many of the processes contributing to the services these ecosystems provide. Such knowledge is essential as many of these ecosystems are in a continued state of global decline. For example, at least 35% of mangrove area was lost between the 1970s and 2000 with annual loss rates of 1–2% (Valiela et al. 2001). Concurrently, indirect habitat alterations degrade ecological services of mangrove systems. While the effects of the loss and degradation of services such as coastal protection, carbon sequestration, and the provision of raw materials are more easily quantified, the effect of lost coastal ecosystem function on the maintenance of marine and estuarine communities is not easily delineated (Barbier et al. 2011).

A meta-analysis of long-term data on coastal ecosystems revealed severe historical declines in multiple coastal ecosystem services, including a 33% decline in the number of viable fisheries and a 69% loss of nursery habitat (Worm et al. 2006). These two services are intimately linked since nurseries supply new recruits to support exploited fisheries. In general, coastal nurseries (e.g., mangroves) are believed to increase survivorship of young fish and thereby influence community structure and increase biomass on connected adult habitats (Dahlgren and Eggleston 2000; Grol et al. 2011; Barbour 2013). This enhancement is important in the support of fisheries through increased recruitment, as well as for maintaining natural ecosystem structure. For example, a nursery-enhancing requirement for species that control macroalgal growth on coral reefs protects against phase shifts from coral to macroalgal reefs.

Recently, much attention has been directed towards defining what constitutes nursery habitat in order to improve our ability to understand and identify such areas. In general, nursery habitats are defined as juvenile habitats making a disproportionate contribution to the adult population, either on a per unit area basis or through total contribution. Since such contribution is difficult to measure directly, studies have typically focused on quantifying juvenile density, growth, survival, and/or connectiv-

* Corresponding author: ufsnook@gmail.com

ity to adult habitats (Dahlgren and Eggleston 2000; Grol et al. 2011; Barbour 2013). These measurements allow a comparison of potential nursery value between habitats but only identify the end result of the processes occurring within those habitats. Without an understanding of time- and size-dependent processes contributing to the nursery value of said habitat, future management of the nursery service of coastal ecosystems will be ineffective.

These processes involve, but are not limited to, the potential importance of multiple habitats in contributing to an overall nursery habitat function and to the mechanistic characteristics contributing to nursery habitat quality (e.g., abundant food, predation refuge). For example, Dahlgren and Eggleston (2000) found an important trade-off between growth and mortality that was reliant upon body size and resulted in a series of ontogenetic shifts within the juvenile phase of Nassau Grouper Epinephelus striatus. If their study did not measure multiple metrics over multiple body sizes, they may have misidentified the nursery value of algal and postalgal habitats, which act in concert to create a nursery mosaic over multiple sizes. Furthering this concept in order to explore the mechanistic characteristics of the growth to mortality ratio, Grol et al. (2011) explored this ratio in the French Grunt Haemulon flavolineatum between reef and nonreef habitats. They concluded that although there was a trade-off between maximizing growth and minimizing mortality, predation risk appeared to be the primary driver for size-based distribution across ecosystems.

Such insights into the processes and mechanisms driving nursery habitat function allow for creation of more effective management practices that conserve the mechanisms responsible for this ecosystem service. For example, mangrove creeks are a putative nursery habitat for multiple species. Although the mangrove shorelines of Charlotte Harbor, Florida have been directly protected for three decades, the functionality of the creek's nursery service has been degraded by continued upland watershed development, which is known to reduce prey diversity. This example demonstrates that direct protection of a nursery habitat without regard to the processes contributing to its value is an incomplete management practice. To improve our ability to manage the juvenile phase of marine and estuarine fishes, and thereby improve stability of connected communities, we need to form a better understanding of the mechanisms driving the nursery function of coastal ecosystems in replenishing adult populations.

In this work, we described habitat use of a juvenile, mangrove-dependent piscivore in order to improve understanding of the nursery function of coastal ecosystems. We conducted a mark–resighting study of Common Snook *Centropomus undecimalis* in Charlotte Harbor, Florida, USA. During a 3-year study, we marked 1,917 juveniles (120–320 mm standard length) with passive integrated transponder (PIT) tags, resighting 85.7% with a PIT tag antenna array in four mangrove creeks (full details in Barbour and Adams 2012; Barbour 2013). Tagging mainly occurred from November through February. Emigration rates and survival varied seasonally with lowest emigration rates in the coldest months and lowest apparent survival in the spring. Emigration events during the fall and winter tended to be temporary movements while emigration in the spring to summer resulted in more permanent emigration. We interpret this as a potential seasonal transition to alternative habitats, with increased mortality at the end of age 0 acting as a potential early population bottleneck (Barbour 2013). Additionally, we found that anthropogenic habitat alterations may act as a thermal refuge during thermal disturbance events and that juveniles spatially segregate from adults to reduce intercohort cannibalism, which is supported by discrete seine-net data (Barbour 2013). Additionally, low apparent survival estimate values for the first-capture interval suggest the presence of transient individuals, which could be explained through an intrapopulation variation in movement tactic wherein some individuals are more likely to move than others (Barbour 2013).

This work revealed that individual movement within an overall nursery mosaic, with seasonal changes in nursery use, is a valid early life history strategy for estuarine fishes. We suggest that mangrove creeks can function as part of a habitat mosaic with bidirectional movements being common and seasonal nursery valuation being necessary. Future fish studies in mangrove creeks should consider the role of seasonal dynamics and multidirectional movements while identifying the casual drivers of movement. These findings have direct implications for Common Snook management and habitat conservation and suggest that work on other species might be reevaluated to determine if movement (rather than mortality) is a more important factor in the juvenile life stage than previously assumed.

Acknowledgments

Funding was provided by a National Science Foundation Graduate Research Fellowship under Grant No. DGE-0802270, the National Sea Grant College Program of the U.S. Department of Commerce's National Oceanic and Atmospheric Administration under Grant No. NA06OAR4170014, Florida Fish and Wildlife Conservation Commission Project No. 11409, the Southwest Florida Water Management District. We thank T. Yess, R. Boucek, J. Adams, and numerous volunteers for assistance in the field, and D. and D. Gulnac for facilities.

References

Barbier, E. B., S. D. Hacker, C. Kennedy, E. W. Koch, A. C. Stier, and B. R. Silliman. 2011. The value of estuarine and coastal ecosystem services. Ecological Monographs 81:169–193.

Barbour A. B. 2013. Processes affecting nursery habitat value of an estuarine-dependent fish. Doctoral dissertation. University of Florida, Gainesville.

Barbour, A. B., and A. J. Adams. 2012. Biologging to examine multiple life stages of an estuarine-dependent fish. Marine Ecology Progress Series 457:241–250.

Dahlgren, C. P., and D. B. Eggleston. 2000. Ecological processes underlying ontogenetic habitat shifts in a coral reef fish. Ecology 81:2227–2240.

Grol, M. G., I. Nagelkerken, A. L. Rypel, and C. A. Layman. 2011. Simple ecological trade-offs give rise to emergent cross-ecosystem distributions of a coral reef fish. Oecologia 165:79–88.

Valiela, I., J. L. Bowen, and J. K. York. 2001. Mangrove forests: one of the world's threatened major tropical environments. BioScience 51:807–815.

Worm, B., E. B. Barbier, N. Beaumont, J. E. Duffy, C. Folke, B. S. Halpern, J. B. C. Jackson, H. K Lotze, F. Micheli, S. R. Palumbi, E. Sala, K. A. Selkoe, J. J. Stachowicz, and R. Watson. 2006. Impacts of biodiversity loss on ocean ecosystem services. Science 314:787–790.

Potential Impacts from Global Climate Change and Other Disturbances

American Fisheries Society Symposium 83:129–131, 2015

EXTENDED ABSTRACT

Sea Level Rise Impacts on a Mangrove Ecosystem in Indaramayu, West Java: Implications on Mitigation and Adaptation

Agustinus M. Samosir*, Sulistio Sulistiono, and Matt F. Rahardjo
Department of Aquatic Resources Management, Bogor Agricultural University
Jl. Agatis, Kampus IPB Dramaga, Bogor 16680, Indonesia

Mangroves in Java have been experiencing significant loss and degradation in the past 25 years; after being extensively cleared and misused as shrimp ponds in the early 1990s, only a few patches of mangrove are left. Some of these remaining mangrove areas have recovered naturally or by rehabilitation programs; however, another threat is impeding the recovery of mangroves. Climate change has been causing devastating damage to the mangrove ecosystems. The damage is from rising sea levels, increased intensity of rain and drought, and acceleration of storms (Ellison 2012). This work aims to identify local vulnerability factors, assess the impact on structures and function of mangrove ecosystems, and formulate adaptation measures for rising sea levels.

Observations were carried out monthly for 6 months in the deltas of Cimanuk River branches in Indramayu, West Java, to see the ecological impact of rising sea level (the area was experiencing long floods in 2008 and 2014) on mangroves and fishes. Two subdeltas were selected to represent an eroded and uneroded mangrove ecosystem, Pabean Ilir (uneroded) and Pagirikan (eroded). Both mangroves are on the Cimanuk delta, have similar land use, and differ only in level of abrasion. The mangroves were further divided into fringing shallow sea, riverine, and fish pond subhabitats. Three locally used types of fishing equipment were employed for sampling each habitat: gill net for the fringing shallow sea, set net for the river, and bamboo fish traps for the fish pond. Diversity and secondary productivity of mangroves and fishes were assessed individually in each subhabitat to see the effect of rising sea level on the structure and function of these ecosystems (Faunce and Serafy 2006). The productivities were classified into three categories: all fish communities, the finfish community, and the dominant fish population. Mangrove vegetation data that were collected included vegetation type, frequency, density, and stand dominance. The measurements were collected within 10 × 10 m^2 within each subhabitat. For environmental conditions, water quality was collected from each site, focusing on the key parameters, primary production (chlorophyll content), and salinity. Analysis of variance was utilized to see the differences and similarities within the variables, and a Marxan model was used to formulate alternative conservation zones and measures. These management measures should increase the resilience of the delta to the impact of rising sea levels.

* Corresponding author: agus.samosir@gmail.com

It was observed that rising tide, flood, storm, and saltwater intrusion were the major sea level factors affecting the Indramayu mangrove ecosystems. The different impacts between the Pabean Ilir and Pagirikan subdeltas of the Cimanuk were represented by mangrove condition and water quality. The density of mangrove trees in Pabean Ilir ranged from 10 to 310 per 100 m^2, while in Pagirikan, density ranged from 1 to 165 per 100 m^2. Six true species of mangrove were found in Pabean Ilir, with the predominant species being red mangrove *Rhizophora mucronata*, and five species were found in Pagirikan, which was dominated by red mangrove and *Avicenia alba*. In terms of water quality, average chlorophyll concentration in Pabean Ilir and Pagirikan was 23.4 and 11.8 4 μg/L, respectively. The salinity was 23.3–36 parts per thousand in Pabean Ilir and 24–30 parts per thousand in Pagirikan. In both cases, average salinity in shallow sea was greater than the river mouth, which was greater than the fish pond. Shannon diversity index (based on weight $\log^2$) of fishes in Pabean Ilir was 2.85 compared to 2.10 in Pagirikan.

The predominant species in Pabean was Glass Perch *Ambassis vachelii*, while in Pagrikan, it was Common Ponyfish *Leiognathus equulus*. The lowest fish diversity was found in the fish ponds, followed by river mouths and then shallow sea. The low fish diversity in the fish ponds was expected as they are enclosed for the culture of Milkfish *Chanos chanos*. Bonefish pesticide is used to eliminate other species of fish in the ponds. In 2013, a total of 180.341 metric tons of finfish and shellfish were produced annually in the Pabean Ilir subdelta, almost the same as in Pagirikan (175.284 metric tons per year); however, in terms of productivity, Pabean Ilir produces 401 $g \cdot m^{-2} \cdot year^{-1}$, much higher than Pagirikan, which produces 204 $g \cdot m^{-2} \cdot year^{-1}$. In Pabean Ilir, the productivity predominantly comes from the creek mouth, while in Pagirikan, it comes from the shallow sea. Secondary productivity of Common Ponyfish at both sites showed similar trends; the productivity/biomass ratio of Common Ponyfish was 3.8 per year in Pabean Ilir and 4.2 in Pagirikan.

The analysis of the ecology of an eroded and uneroded mangrove ecosystem on the deltas of the Cimanuk River showed that the protected (uneroded) mangrove had higher mangrove cover, biodiversity, and secondary productivity compared to that of the open coast. The protected mangrove also had lower abrasion, saltwater intrusion, and secondary production.

Marxan (spatial conservation annealing) software (Steward and Possingham 2005) was used to develop strategies to increase the ecosystem resiliency of the mangroves. A design of conservation area was proposed, which consisted of core, limited use and sustainable fishery zones (Figure 1). Five programs were recommended to be implemented in the three zones of the conservation area: greenbelt establishment and mangrove plantating in the eroded core zone, recreational fishing regulation in the limited-used zone, and fish size control and mariculture development in the sustainable fishery zone.

References

Ellison, J. C. 2012. Climate change vulnerability assessment and adaptation planning for mangrove systems. World Wildlife Fund, Washington, D.C.

Faunce, C. H., and J. E. Serafy. 2006. Mangroves as fish habitat: 50 years of field studies. Marine Ecology Progress Series 318:1–18.

Steward, R. R., and H. P. Possingham. 2005. Efficiency, costs, and trade-off in marine reserve system design. Environmental Modelling and Assessment 10:203–213.

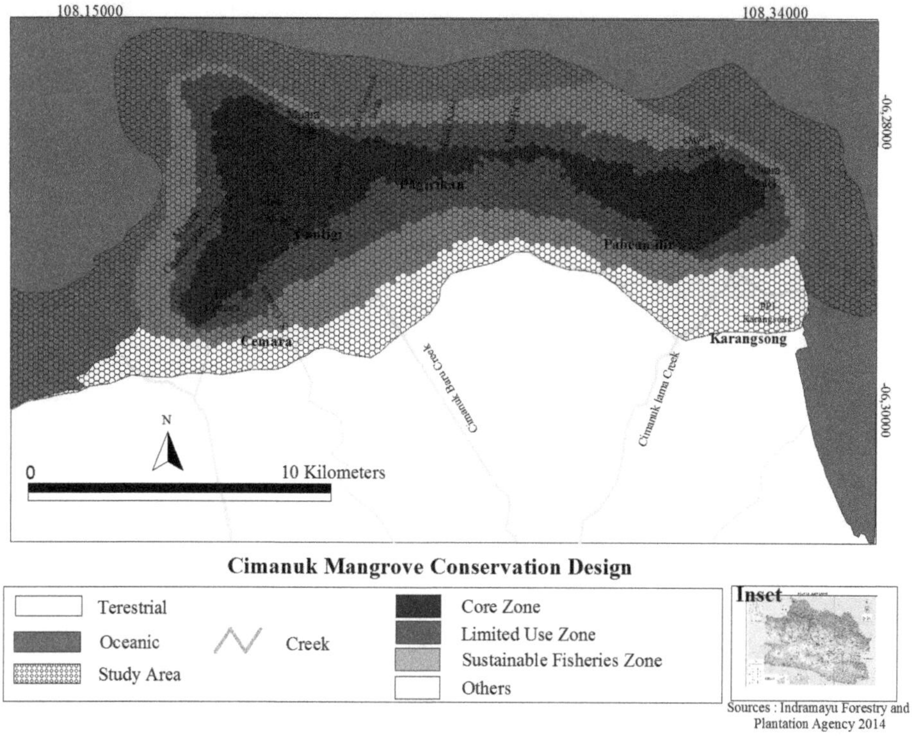

Figure 1.—The proposed design for mangrove fishery conservation in the delta.

Mangrove Restoration Success Stories

American Fisheries Society Symposium 83:135–143, 2015

EXTENDED ABSTRACT

Colonization of Robinson Preserve: An Evaluation of Restoration Efforts in an Estuary in Tampa Bay, Florida

Amanda Croteau*
Program in Fisheries and Aquatic Sciences, University of Florida
7922 NW 71st Street, Gainesville, Florida 32653, USA

Mangrove forests are ecologically important coastal ecosystems found throughout much of the tropics and subtropics. Worldwide, mangroves are experiencing losses of 1–2% per year (Lewis 2005). Many of Florida's mangrove habitats, including those in Tampa Bay, have been severely impacted by development with statewide losses of more than 75% (Lewis 2005). Such habitat loss can have major impacts on local fisheries. In Florida, 80% of all commercially or recreationally targeted marine species depend upon mangrove estuarine areas during some stage of their life cycle (Lewis et al. 1985). Robinson Preserve is a 197-ha preserve and one of the largest salt marsh and mangrove restoration efforts in Tampa Bay. Originally a coastal wetland, the property was ditched, drained, and used for agriculture. In 2006, reconnecting the property's existing water bodies to the Manatee River, Perico Bayou, and Palma Sola Bay restored tidal flow. Existing old growth mangroves were preserved and new mangroves were expected to colonize in areas with high propagule pressure (i.e., areas where large amounts of mangrove seeds/seedlings are frequently deposited by currents and tides). Upland and salt marsh vegetation were planted; however, aquatic flora and fauna were left to colonize from neighboring populations.

Evaluation of coastal restoration efforts often focuses on vegetation (coverage, density, and structure, etc.), and the time of evaluation is usually shorter than the time required for function to be restored (Simenstad and Thom 1996). Lewis and Gilmore (2007) noted that fish populations within restored marshes and mangroves may recover to levels similar to reference sites in both species composition and density within 5 years. The waters of Robinson Preserve were sampled seasonally from 2008 to 2013 to evaluate the success of restoration activities, and colonization by fish. The preserve was divided into four regions based on water flow and connectivity to surrounding water bodies (1: mixing zone, 2: Palma Sola Bay and Perico Bayou, 3: freshwater upland inputs, and 4: Manatee River). Sampling was distributed throughout each zone with effort proportional to zone size. Each zone was sampled using multiple gears (21.3-m center-bag seine, 6-m beach seine, 2.1-m cast net, and a 0.36-m bottom frame dip net) to effectively target species of varying sizes and habitat preferences. The proportional effort of each gear was similar among the different zones. Sampled habitats were classified by shore type.

A preliminary study conducted 1 year after tidal reconnection found 18 fish spe-

* Corresponding author: acroteau@ufl.edu

cies (Croteau 2007). At the conclusion of this study, 7 years after tidal reconnection, 85 species of fish were documented in Robinson Preserve. Numerous commercially and recreationally important species utilize the preserve as a nursery for juveniles and/or as adult foraging (e.g., Common Snook *Centropomus undecimalis*, Red Drum *Sciaenops ocellatus*, Atlantic Tarpon *Megalops atlanticus*, and Striped Mullet *Mugil cephalus)*. Fish species richness differed by season (analysis of variance, $P < 0.0015$). Greater numbers of fish species were collected during the spring, summer, and fall compared to winter, which had the lowest number of species on average. Some of the species captured are seasonal either due to migratory patterns or the timing of juvenile recruitment.

Species were ranked by the total number caught throughout the sampling period for locations with and without mangroves present along the shoreline being sampled. In general, there was a high degree of overlap between these two habitat classifications (Table 1). Ninety-five percent of the fish species documented in Robinson Preserve were found in nonmangrove locations compared to 73% found in association with mangroves (Figure 1). However, the species that were not collected at mangrove sites during the course of this study were primarily rare species (few individuals collected and/or observed throughout the study) and were often benthic, pelagic, or freshwater associated species, which would not be expected to preferentially associate with mangrove-dominated shorelines. Some of these species may have been documented in nonmangrove sites simply due to the greater extent and sampling of such areas. Several species, such as Common Snook and Red Drum, were found in both habitat types but typically ranked higher (i.e., were found in greater numbers relative to other species documented in the same habitat type) in mangrove sites than nonmangrove sites.

Cumulative species curves were developed for the preserve as a whole and each of the regions, with number of species documented to date for all habitat types, mangrove sites, and nonmangrove sites (Figure 1). Although new species were documented in the last sampling event, the cumulative species curves have approached asymptotes (<5% change over the past 10 sampling events). Within the four sampling regions, mangrove sites had fewer numbers of new species per event, except for zone 4, the region with the most intact old-growth mangroves. Since 2011, more than 80% of the total species colonizing that zone have been associated with mangrove habitats.

Robinson Preserve is still in transition, as vegetation grows in coverage area and density. Few differences in mangrove and nonmangrove habitats were observed, except in the zone with the largest percent coverage of old-growth mangroves (zone 4). Young mangroves may not supply the same habitat quality as older mature stands of mangroves (Bosire et al. 2008), although they do appear to be serving as important habitat within the restored ecosystem. The data utilized in this study came from a larger project whose goal was to evaluate the overall restoration of the preserve, not to compare mangrove to nonmangrove communities. The majority of the habitats within the preserve are dominated by saltmarsh vegetation, not mangroves, so sampling of mangrove habitats was restricted in each zone due to relative distribution of mangrove-dominated shorelines. More directed sampling in mangrove habitats within the preserve, particularly old-growth mangroves, may have revealed greater usage of mangrove by some species or more species that are unique to mangrove habitats. However, even with such potential biases, it is clear that mangroves are serving as important fish habitat to a number of species, including several species with targeted fisheries. More detailed analyses of

Table 1.—Top 25 ranked species by total number caught in both mangrove and nonmangrove habitats for all zones combined and each individual zone. Bold denotes a recreationally or commercially important species.

All zones				Zone 1			
Mangrove		Nonmangrove		Mangrove		Nonmangrove	
Species	Rank	Species	Rank	Species	Rank	Species	Rank
Bay Anchovy *Anchoa mitchilli*	1	Bay Anchovy	1	Bay Anchovy	1	Bay Anchovy	1
mojarra *Eucinostomus* spp.	2	Mojarra	2	**Spot**	**2**	Mojarra	2
Spot *Leiostomus xanthurus*	**3**	Tidewater Silverside	3	Mojarra	3	**Pinfish**	**3**
Tidewater Silverside *Menidia peninsulae*	4	**Spot**	**4**	Tidewater Silverside	4	Tidewater Silverside	4
Gulf Menhaden *Brevoortia patronus*/Yellowfin Menhaden *B. smithi*	**5**	**Pinfish**	**5**	**Gulf Menhaden/ Yellowfin Menhaden**	5	**Spot**	**5**
Longnose Killifish *Fundulus similis*	6	Rainwater Killifish	6	Longnose Killifish	6	**Gulf Menhaden/ Yellowfin Menhaden**	6
Sheepshead Minnow *Cyprinodon variegatus*	7	**Gulf Menhaden/ Yellowfin Menhaden**	**7**	Sheepshead Minnow	7	Longnose Killifish	7
Pinfish *Lagodon rhomboides*	**8**	Sheepshead Minnow	8	Gulf Killifish	8	**Scaled Sardine**	**8**
Gulf Killifish *F. grandis*	9	Longnose Killifish	9	Diamond Killifish	9	Gulf Killifish	9
Diamond Killifish *Adinia xenica*	10	Gulf Killifish	10	**Pinfish**	**10**	**Striped Mullet**	**10**
Sailfin Molly *Poecilia latipinna*	11	Sailfin Molly	11	Sailfin Molly	11	Rainwater Killifish	11
Clown Goby *Microgobius gulosus*	12	**Scaled Sardine**	**12**	Striped Mojarra	12	Sheepshead Minnow *Cyprinodon variegatus*	12
Striped Mojarra *Eugerres plumieri*	13	**Striped Mullet**	**13**	**White Mullet/ Fantail Mullet**	**13**	Sailfin Molly	13
Striped Mullet *Mugil cephalus*	**14**	**Common Snook**	**14**	**Striped Mullet**	**14**	goby *Gobiosoma* spp.	14
White Mullet *M. curema*/ Fantail Mullet *M. trichocdon*	**15**	Clown Goby	15	Marsh Killifish	15	Clown Goby	15
Marsh Killifish *F. confluentus*	16	Striped Mojarra	16	Clown Goby	16	Striped Mojarra	16
Common Snook *Centropomus undecimalis*	**17**	Silver Perch *Bairdiella chrysoura*	17	**Red Drum**	**17**	Diamond Killifish	17
Rainwater Killifish *Lucania parva*	18	goby *Gobiosoma* spp.	18	**Common Snook**	**18**	Silver Perch	18
Scaled Sardine *Harengula jaguana*	**19**	Diamond Killifish	19	Rainwater Killifish	19	**White Mullet/ Fantail Mullet**	19

Table 1.—Continued.

All zones				Zone 1			
Mangrove		Nonmangrove		Mangrove		Nonmangrove	
Species	Rank	Species	Rank	Species	Rank	Species	Rank
Sheepshead *Archosargus probatocephalus*	**20**	**Sheepshead**	**20**	**Ladyfish**	**20**	**Common Snook**	**20**
Red Drum *Sciaenops ocellatus*	**21**	**White Mullet/ Fantail Mullet**	**21**	Blackchin Tilapia	21	Pigfish *Orthopristis chrysoptera*	21
Ladyfish *Elops saurus*	**22**	Lined Sole	22	Leatherjack *Oligoplites saurus*	22	**Sheepshead**	**22**
Blackchin Tilapia *Sarotherodon melanotheron*	23	Redfin Needlefish *Strongylura notata*	23	Lined Sole	23	Gulf Pipefish	23
Goldspotted Killifish *Floridichthys carpio*	24	Gulf Pipefish *Syngnathus scovelli*	24	Redfin Needlefish	24	**Red Drum**	**24**
Lined Sole *Achirus lineatus*	24	**Red Drum**	**25**	**Spotted Seatrout *Cynoscion nebulosus***	**25**	Leatherjack	25

Zone 2				Zone 3			
Mangrove		Nonmangrove		Mangrove		Nonmangrove	
Species	Rank	Species	Rank	Species	Rank	Species	Rank
Bay Anchovy	1	Bay Anchovy	1	Mojarra	1	Bay Anchovy	1
Mojarra	2	Mojarra	2	**Gulf Menhaden/ Yellowfin Menhaden**	**2**	Rainwater Killifish	2
Spot	**3**	**Spot**	**3**	Bay Anchovy	3	Mojarra	3
Gulf Menhaden/Yellowfin Menhaden	**4**	Tidewater Silverside	4	**Pinfish**	**4**	Tidewater Silverside	4
Pinfish	**5**	**Pinfish**	**5**	**Sheepshead**	**5**	**Spot**	**5**
Clown Goby	6	**Gulf Menhaden/ Yellowfin Menhaden**	**6**	**Scaled Sardine**	**6**	**Gulf Menhaden/ Yellowfin Menhaden**	**6**
Sailfin Molly	7	Sheepshead Minnow	7	Striped Mojarra	7	**Common Snook**	**7**
Gulf Killifish	8	Rainwater Killifish	8	**Spot**	**8**	**Pinfish**	**8**
Tidewater Silverside	9	Longnose Killifish	9	**Striped Mullet**	**9**	Sheepshead Minnow	9
Longnose Killifish	10	Clown Goby	10	**White Mullet/Fantail Mullet**	**10**	Sailfin Molly	10
Sheepshead Minnow	11	Gulf Killifish	11	**Common Snook**	**11**	Gulf Killifish	11

Table 1.—Continued.

Zone 2				Zone 3			
Mangrove		Nonmangrove		Mangrove		Nonmangrove	
Species	Rank	Species	Rank	Species	Rank	Species	Rank
Common Snook	**12**	Sailfin Molly	12	**Gray Snapper**	**12**	**Striped Mojarra**	**12**
Striped Mullet	**13**	Silver Perch	13	Blackchin Tilapia	13	Longnose Killifish	13
Goldspotted Killifish	14	**Striped Mullet**	**14**	Tidewater Silverside	13	Striped Mullet	14
Sheepshead	**15**	Striped Mojarra	15	Sailfin Molly	15	Clown Goby	15
Rainwater Killifish	16	Goldspotted Killifish	16	**Ladyfish**	**16**	Western Mosquitofish *Gambusia affinis*	16
Diamond Killifish	17	goby *Gobiosoma* spp.	17	**Great Barracuda *Sphyraena barracuda***	**17**	**Sheepshead**	**17**
Striped Mojarra	18	Redfin Needlefish	18	Redfin Needlefish	17	Blackchin Tilapia	18
Lined Sole	19	**Common Snook**	**19**	**Black Drum *Pogonias cromis***	**19**	goby *Gobiosoma* spp.	19
Red Drum	**20**	Lined Sole	20	Gulf Killifish	19	Lined Sole	19
Southern Puffer *Sphoeroides nephelus*	20	**Sheepshead**	**21**	**Red Drum**	**19**	**Red Drum**	**19**
Scaled Sardine	**22**	**Red Drum**	**22**	**Spotted Seatrout**	**19**	Gulf Pipefish	22
White Mullet/Fantail Mullet	**23**	**Scaled Sardine**	**23**	Bighead Searobin *Prionotus tribulus*	23	**Gray Snapper**	**23**
Gray Snapper *Lutjanus griseus*	**24**	**White Mullet/Fantail Mullet**	**24**	**Crevelle Jack *Caranx hippos***	**23**	Blue Tilapia *Oreochromis aureus*	24
Frillfin Goby *Bathygobius soporator*	25	Gulf Toadfish *Opsanus beta*	25	**Gulf Flounder *Paralichthys albigutta***	**23**	Diamond Killifish	25
				Leatherjack	23	**Ladyfish**	**25**
				Longnose Killifish	23	**Scaled Sardine**	**25**
				Rainwater Killifish	23		
				Silver Perch	23		

Table 1.—Continued.

Zone 4			
Mangrove		Nonmangrove	
Species	Rank	Species	Rank
Bay Anchovy	1	Bay Anchovy	1
Mojarra	2	Mojarra	2
Tidewater Silverside	3	**Pinfish**	**3**
Pinfish	**4**	Tidewater Silverside	4
Spot	**5**	**Scaled Sardine**	**5**
Scaled Sardine	**6**	Longnose Killifish	6
Gulf Killifish	7	Gulf Killifish	7
Sailfin Molly	8	Sailfin Molly	8
Rainwater Killifish	9	**Spot**	**9**
Clown Goby	10	Silver Perch	10
Gulf Menhaden/Yellowfin Menhaden	**11**	**Gulf Menhaden/ Yellowfin Menhaden**	**11**
White Mullet/Fantail Mullet	**12**	Sheepshead Minnow	12
Common Snook	**13**	**Atlantic Thread Herring**	**13**
goby *Gobiosoma* spp.	14	Rainwater Killifish	14
Redfin Needlefish	15	Redfin Needlefish	15
Blackchin Tilapia	16	Gulf Pipefish	16
Longnose Killifish	17	Southern Puffer	17
Striped Mojarra	18	Blackchin Tilapia	18
Leatherjack	19	Lined Sole	18
Sheepshead Minnow	19	Striped Mojarra	18
Atlantic Thread Herring *Opisthonema oglinum*	**21**	goby *Gobiosoma* spp.	21
Sheepshead	**22**	**Clown Goby**	22
Lined Sole	23	**Common Snook**	**22**
Silver Perch	24	**Striped Mullet**	**24**
Goldspotted Killifish	25	Leatherjack	25

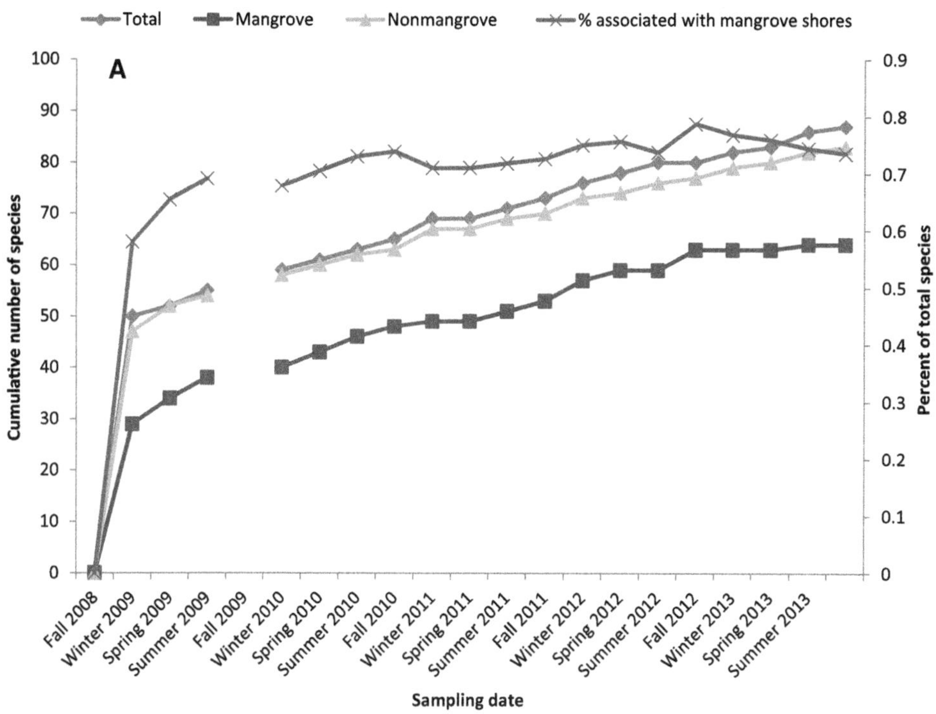

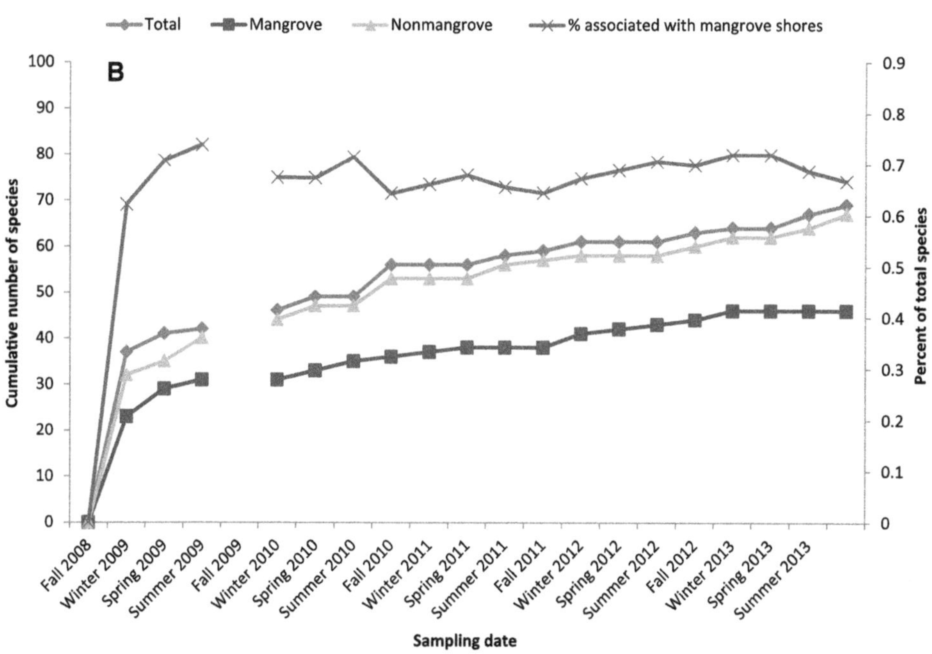

Figure 1.—Cumulative species curves of (A) Robinson Preserve, and each of the four zones determined by water flow and connectivity to neighboring water bodies: (B) Zone 1: mixing zone, (C) Zone 2: Palma Sola Bay and Perico Bayou (D) Zone 3: upland freshwater drainage, and (E) Zone 4: Manatee River. Sampling was conducted fall 2008 through summer 2013, with the exception of summer 2009.

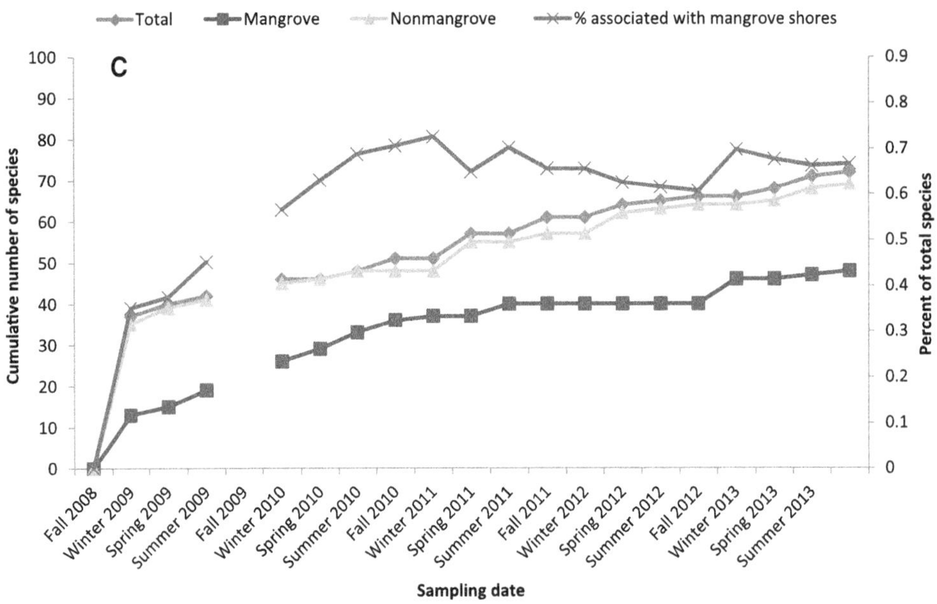

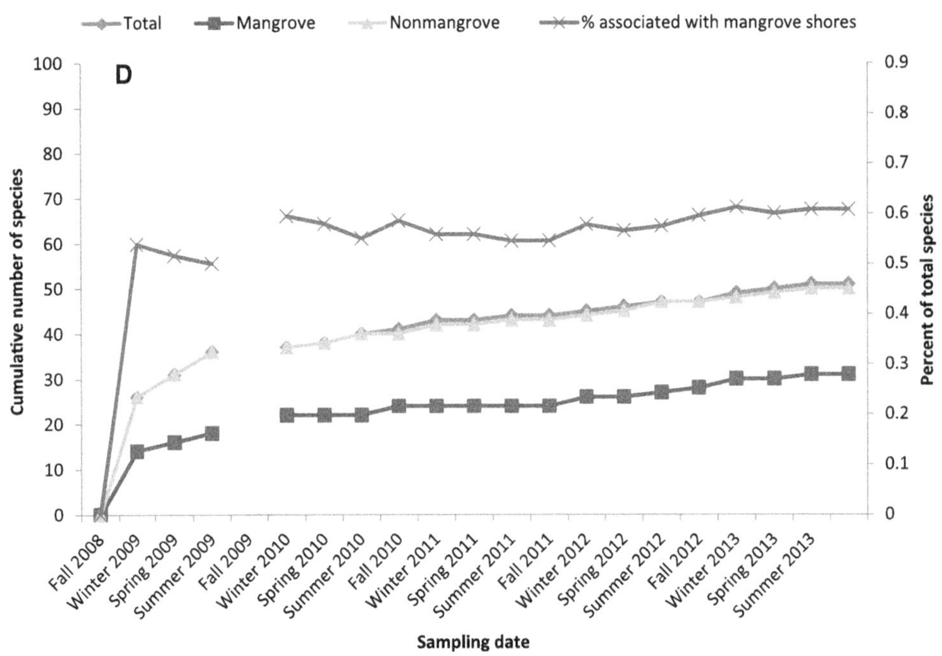

Figure 1.—Continued.

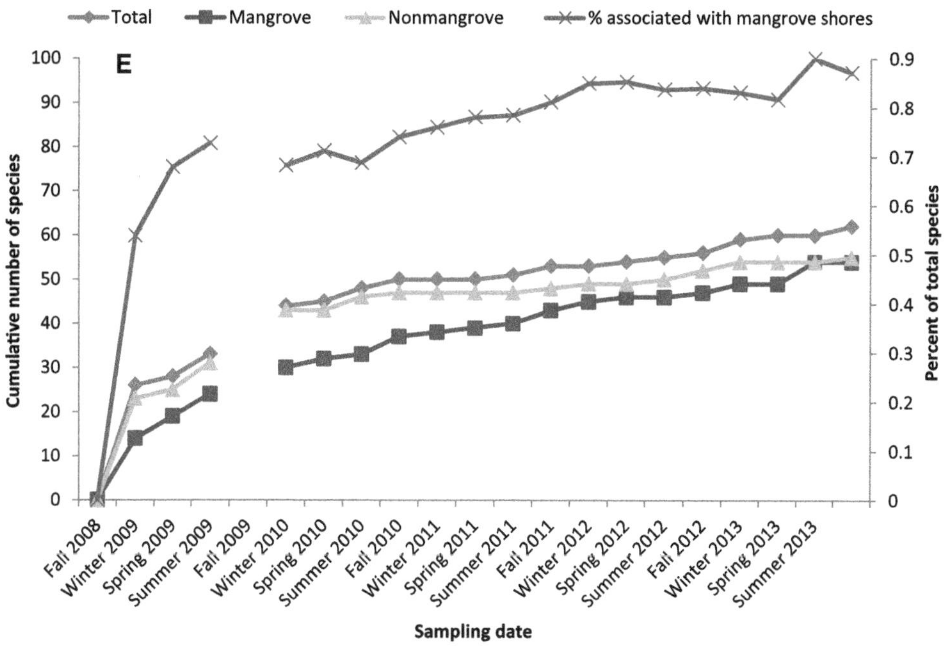

Figure 1.—Continued.

the habitat data (e.g., separating the nonmangrove habitat into different groups), which will be completed in the future, may also reveal a greater importance of mangroves to some species.

References

Bosire, J. O., F. Dahdouh-Guebas, M. Walton, B. I. Crona, R. R. Lewis, III, C. Field, J. G. Kairo, and N. Koedam. 2008. Functionality of restored mangroves: a review. Aquatic Botany 89:251–259.

Croteau, A. 2007. Robinson Preserve: aquatic survey to assess restoration and recruitment (October 2007). Manatee County Natural Resources Department, Quarterly Report, Bradenton, Florida.

Lewis, R. R. 2005. Ecological engineering for successful management and restoration of mangrove forests. Ecological Engineering 24:403–418.

Lewis, R. R., and R. G. Gilmore. 2007. Important considerations to achieve successful mangrove forest restoration with optimum fish habitat. Bulletin of Marine Science 80:823–837.

Lewis, R. R., R. G. Gilmore, Jr., D. W. Crewz, and W. E. Odum. 1985. Mangrove habitat and fishery resources of Florida. Pages 281–336 *in* W. Seaman, editor. Florida aquatic habitat and fishery resources. American Fisheries Society, Florida Chapter, Kissimmee, Florida.

Simenstad, C. A., and R. M. Thom. 1996. Functional equivalency trajectories of the restored Gog-Le-Hi-Te estuarine wetland. Ecological Applications 6:38–56.

American Fisheries Society Symposium 83:145–148, 2015

EXTENDED ABSTRACT

Ichthyoplankton and Meroplankton Community Composition of a Restored Mangrove and Salt Marsh Aquatic Preserve in Tampa Bay, Florida

AMANDA CROTEAU*
Program in Fisheries and Aquatic Sciences, University of Florida
7922 NW 71st St, Gainesville, Florida 32653, USA

Mangroves and salt marshes provide vital juvenile habitat for many nearshore and offshore marine species. Florida's coastal habitats have been severely impacted by coastal development, and Tampa Bay has lost more than 44% of its mangrove and salt marsh habitats (Lewis et al. 1985). Robinson Preserve is one of the largest (197 ha) mangrove and salt marsh restoration efforts in Tampa Bay. Robinson Preserve was originally a coastal wetland that was ditched and drained in the 1920s for agricultural use. In 2006, tidal flow was restored through connections with Perico Bayou, Palma Sola Bay, and the Manatee River. Restoration also involved the planting of native upland and salt marsh vegetation. However, no efforts were made to supplement the aquatic flora and fauna; rather, it was expected that they would colonize the preserve from neighboring populations. Ichthyoplankton and meroplankton abundances were selected as one metric to evaluate the quality of the restored ecosystem as nursery habitat.

A number of methods exist for sampling ichthyoplankton and other large zooplankton, including light traps, benthic sleds, Miller high-speed samplers, push nets, and tow nets. As with any sampling gear, each method has its pros and cons.

Robinson Preserve is shallow (generally <2 m), with complex habitat types and obstructions. It is also a no motor zone. Due to these study site constraints, light traps were selected as the most efficient gear to sample ichthyoplankton and meroplankton within the preserve. The light trap designed by Jones (2006) was redesigned for deployment from shore and scaled down for use in shallow, estuarine systems.

Robinson Preserve was divided into four regions based on water flow and connectivity to surrounding water bodies (1: mixing zone, 2: Palma Sola Bay and Perico Bayou, 3: uplands freshwater drainage, and 4: Manatee River), with one light trap station per region. Sampling occurred quarterly from 2009 to 2013, beginning 3 years after restoration and tidal reconnection. Sampling dates were chosen to detect possible seasonal changes in community composition (winter: January–February, spring: April–May, summer: July–August, and fall: October–November). Light traps were deployed 8 h before sunrise and retrieved the following morning. Samples were preserved with 5% formalin, filtered, and transferred to 70% ethanol. Ichthyoplankton and other large (≥3 mm) zooplankton were identified and enumerated under a dissecting microscope.

Zooplankton samples have been processed from spring 2009 through spring 2011

* Corresponding author: acroteau@ufl.edu

($N = 30$). To date, 18 major taxonomic groups have been identified. Overall community composition was dominated by isopods (19%), caridean shrimp (18.2%), fish (15.7%), and parasitic copepods (13.1%), though species assemblages varied by site and season (Figure 1). The greatest taxonomic richness was noted in the spring and fall and at sites 1 and 3 (Figure 1). The greater taxonomic richness in the spring and fall could be related to the transition periods between the warmest and coldest periods of water temperatures (i.e., species that prefer warmer and cooler temperatures may be co-occurring in comparison to the summer and winter samples). The greater taxonomic richness in sites 1 and 3 is likely related to their locations. Both of these sites were located in areas with lower current velocities than sites 2 and 4, which may have allowed some less-mobile taxa to enter the light traps than in the latter two sites.

Several fish parasites were collected in large numbers during this study. Juveniles of the parasitic isopod family Cymothidae were by far the most dominant form of isopod present in the light trap samples. They were present in all seasons but had higher abundances in the summer (Figure 1). This may relate to the warmer water temperatures during this season, which have been linked to the reproduction of several parasitic isopod species. Parasitic copepods of the genera *Argulus* and *Caligus* were also collected in large numbers, but unlike the Cymothids, they were most abundant in the winter (Figure 1). Many of the parasitic copepods collected

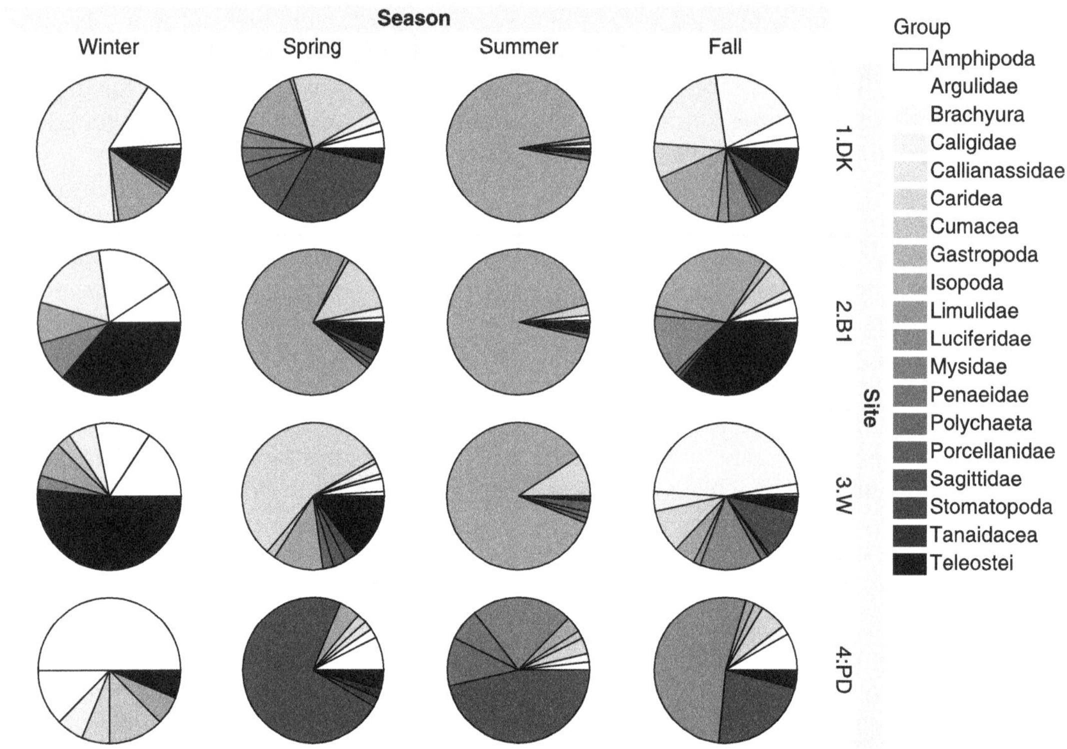

Figure 1.—Ichthyoplankton and meroplankton community composition for each site (1.DK in mixing zone, 2.B1 in Palma Sola Bay and Perico Bayou zone, 3.W in upland freshwater drainage zone, and 4.PD in the Manatee River zone) by season. Samples were analyzed from spring 2009 through spring 2011.

were adults, and their increased abundance in the winter may be related to one or several species of fish migrating into the preserve at this time. One possibility is that fish species, such as Striped Mullet *Mugil cephalus*, that migrate offshore in the fall to spawn may return to the preserve in the winter with a greater load of parasitic copepods. Another possibility is that the parasitic copepods are arriving on adults of some species, such as Common Snook *Centropomus undecimalis*, that utilize the preserve as an overwintering area. The high abundance of external fish parasites collected with this method may provide a new and efficient means of surveying such taxa in estuarine systems.

Larval and settlement stage fish were collected in nearly every sample (Figure 2), including fish from at least eight families (a number of the larval fish have yet to be fully identified). This is similar to the degree of diversity noted in other light trap studies in similar habitats (Hernandez and Shaw 2003; Strydom 2003). Juvenile mullet (likely Striped Mullet) were always present in winter samples while juvenile clupeids (likely menhaden *Brevoortia* spp.) were present in the winter and spring (Figure 2), which corresponds well with their respective peak spawning periods within this region (Lewis and Roithmayr 1981; Mahmoudi 2000). Bay Anchovies *Anchoa mitchilli* were present throughout the year, and based on preliminary examination, a number of the fish larvae that have not been fully identified are likely Bay Anchovies, which spawn year round in Tampa Bay. These three fish taxa are commonly among the most dominant species collected during sampling of juvenile and adult fish within the same preserve.

The use of a modified light trap in Robinson Preserve proved to be an effective method for sampling ichthyoplankton and meroplankton, as well as some other groups. Currently, only the larger (≥3 mm) organisms collected in light trap samples have been pro-

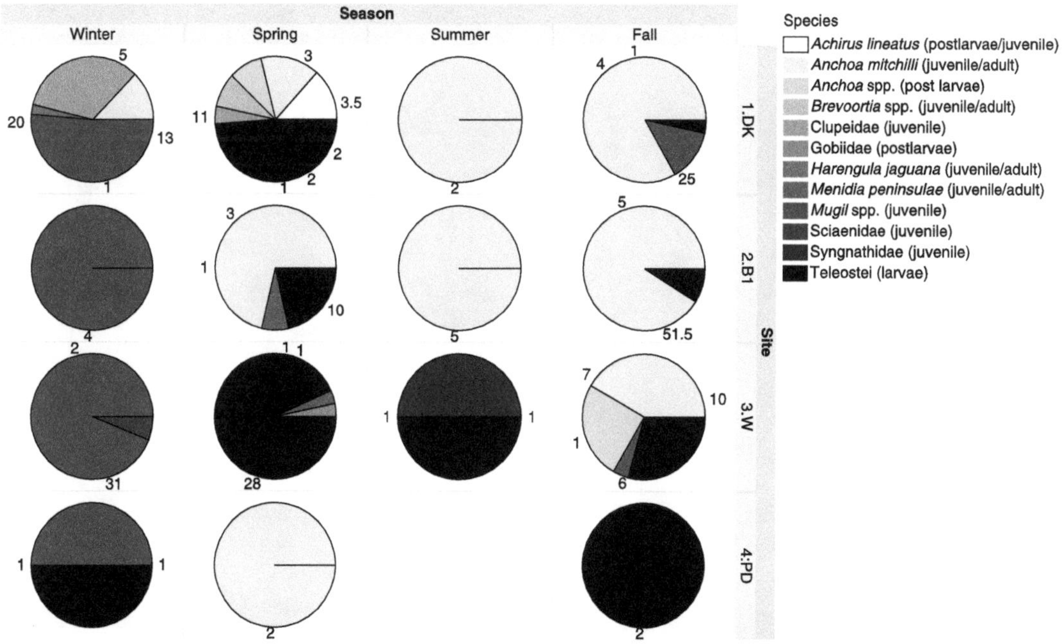

Figure 2.—Ichthyoplankton community composition for each site (1.DK in mixing zone, 2.B1 in Palma Sola Bay and Perico Bayou zone, 3.W in upland freshwater drainage zone, and 4.PD in the Manatee River zone) by season. Samples were analyzed from spring 2009 through spring 2011.

cessed. The identification and enumeration of the smaller organisms along with greater taxonomic resolution of some of the larger taxa will better tie this sampling to concurrent sampling of juveniles and adults within the preserve allowing for a more complete evaluation of restoration efforts.

References

Hernandez, F. J., Jr., and R. F. Shaw. 2003. Comparison of plankton net and light trap methodologies for sampling larval and juvenile fishes at offshore petroleum platforms and a coastal jetty off Louisiana. Pages 15–38 in D. R. Stanley and A. Scarborough-Bull, editors. Fisheries, reefs, and offshore development. American Fisheries Society, Symposium 36, Bethesda, Maryland.

Jones, D. L. 2006. Design, construction, and use of a new light trap for sampling larval coral reef fishes. NOAA Technical Memorandum NMFS-SEFSC-544.

Lewis, R. M., and C. M. Roithmayr. 1981. Spawning and sexual maturity of Gulf Menhaden, Brevoortia patronus. Fishery Bulletin 78:947–951.

Lewis, R. R., R. G. Gilmore, Jr., D. W. Crewz, and W. E. Odum. 1985. Mangrove habitat and fishery resources of Florida. Pages 281–336 in W. Seaman, Jr., editor. Florida aquatic habitat and fishery resources. American Fisheries Society, Florida Chapter, Kissimmee.

Mahmoudi, B. 2000. Status and trends in the Florida mullet fishery and an updated stock assessment. Florida Fish and Wildlife Conservation Commission, IHR 2000 -006, St. Petersburg.

Strydom, N. A. 2003. An assessment of habitat use by larval fishes in a warm temperate estuarine creek using light traps. Estuaries 26:1310–1318.